Handbook of Metal Biotechnology

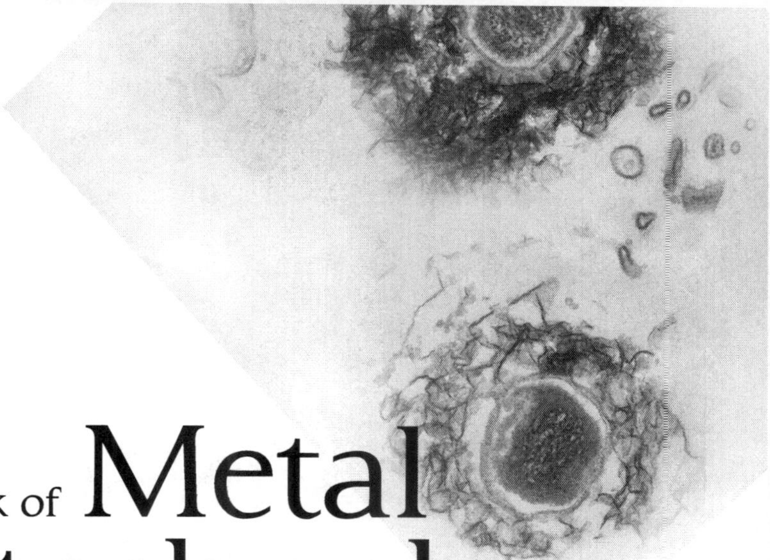

Handbook of Metal Biotechnology

Applications for Environmental Conservation and Sustainability

edited by
Michihiko Ike
Mitsuo Yamashita
Satoshi Soda

PAN STANFORD PUBLISHING

Published by

Pan Stanford Publishing Pte. Ltd.
Penthouse Level, Suntec Tower 3
8 Temasek Boulevard
Singapore 038988

Email: editorial@panstanford.com
Web: www.panstanford.com

British Library Cataloguing-in-Publication Data
A catalogue record for this book is available from the British Library.

Handbook of Metal Biotechnology: Applications for Environmental Conservation and Sustainability

ISBN 978-981-4267-98-4 (Hardcover)
ISBN 978-981-4267-99-1 (eBook)

Printed in the USA

Contents

Preface

Our prosperous life today is highly dependent on the extended use of metals, including rare metals. Copper refining in ca. 5500 BC in Persia is believed to be the oldest use of metal by human beings. After the Bronze and Iron Ages, today all metal elements in the periodical table are indispensable for the modern civilization. These metals are absolutely fundamental to the production of novel materials supplied to high-technology industries.

The sustainable society we aim to establish in the 21st century raises an alarm related to proper use of metal resources in industries and infrastructure. Primarily, metal elements refined from underground resources remain finite and might suffer depletion, presenting society with a serious problem sooner or later. Although we must recycle metal elements to ensure their continued use, complete recycling is practically difficult or impossible because of high cost. Especially it is big challenge to recover metal elements from the drainage, wastes, and off-gas in which they exist in low concentrations and in mixtures of various materials.

Another problem is the environmental contamination by metallic compounds emitted into water bodies, the atmosphere, and soil. These compounds may engender serious health hazards and exert harmful effects on the ecological system, including human beings. Today, rather exotic metals that we have heard of only rarely are slated to be listed as environmental pollutants in addition to well-known toxic heavy metals such as mercury, lead, cadmium, and chromium. Nevertheless, we do not know much about the effective processes and treatments of these new metallic pollutants.

Another alarming concern relates to the attendant energy consumption that is necessary when metal resources are mined, refined, and processed. The physicochemical treatments of these metals consume huge amounts of energy. That consumption itself raises the issue of global warming, which is becoming obvious and is getting worse. It can be said that global warming is the ultimate cost for reaping the social benefits of metals. It must also be remembered that in addition to the enormous consumption of energy, strong acid and alkaline substances are used in metal refining: They are friendly neither to working environments nor to the natural environment.

In the complicated trade-off between metal resources and related issues, namely, resource depletion, environmental pollution, and energy consumption, we propose the development of "metal biotechnology" to cope appropriately with the above-mentioned difficulties. Here, metal biotechnology is defined as biological technology that utilizes the reactions of various metals in metabolism and also chemical binding by living organisms and biomolecules. "Metal" in our metal biotechnology encompasses — in addition to typical metals like iron — all mineral elements, including metalloids such as arsenic and antimony and nonmetals such as selenium and tellurium, which have properties similar to those of typical metal elements.

Biological reactions are intrinsically material saving and energy saving. Therefore, they are both economical and environment friendly. Consequently, the metal biotechnology we propose is a key technology that is useful in solving a problem that no physicochemical process can solve or that no such process can solve easily. Metal biotechnology is expected to offer definite advantages when targeting the reactions in the drainage, wastes, and the environment over physicochemical procedures, because biological metal-related reactions are efficient even when we wish to process metals in low concentrations or in complex compounds. This feature of metal biotechnology will find a wide range of applications in the fields of environmental conservation and metal resource recycling.

This book introduces various fields of metal biotechnology, emphasizing applications for environment conservation and resource recycling. We organized the Metal-Biotechnology Research Division in The Society for Biotechnology, Japan, and solicited contributions for this book to collect 17 monographs. Chapters 1 to 7 address wastewater treatment and bioremediation technologies for hazardous metals making use of metal metabolism by microorganisms and other organisms. Recovery and recycling of metals from drainage and waste sources are discussed in Chapters 8 to 11. Chapters 12 to 14 describe the biological synthesis and processing of new metallic materials and monitoring of metals for industrial uses. In addition to these technologies, topics of bio-informatics in metal biotechnology are included in Chapters 15 to 17. These are expected to be of great help for new developments in these new technologies.

The editors appreciate the efforts by all the authors who contributed their monographs to this publication and thank Mr Stanford Chong, director of Pan Stanford Publishing Pte. Ltd., who recommended that we edit this book.

Editors

Chapter 1

MICROBIAL MANGANESE(II) OXIDATION: A POTENTIAL TOOL FOR TREATMENT OF METAL-CONTAMINATED WATERS

Naoyuki Miyata[1] and Yukinori Tani[2]

[1]*Department of Biological Environment, Akita Prefectural University, Shimoshinjo-Nakano, Akita 010-0195, Japan*

[2]*Institute for Environmental Sciences, University of Shizuoka, Yada 52-1, Shizuoka 422-8526, Japan*

nmiyata@akita-pu.ac.jp; taniy@u-shizuoka-ken.ac.jp

1.1 Introduction

The microbe–metal interactions are considered effective tools for removal of metals from contaminated waters. Direct interactions between microorganisms and metals include the adsorption onto microbial cells, accumulation, and conversion and precipitation driven by microbially mediated reactions (Fig. 1.1). In addition, the mineral precipitates (e.g., Fe and Mn oxides) occurring at microbial cell surfaces often act as adsorbents, oxidants or reductants, and catalysts, owing to their high specific surface area and high reactivity.

Diverse bacteria and fungi oxidize Mn(II) enzymatically and precipitate Mn(III, IV) oxides[1–3] (Fig. 1.1). Under the circumneutral

Handbook of Metal Biotechnology: Applications for Environmental Conservation and Sustainability
Edited by Michihiko Ike, Mitsuo Yamashita, and Satoshi Soda

www.panstanford.com

conditions, the microbial oxidation of Mn(II) proceeds at several orders of magnitude faster than abiotic oxidation. The microbial oxidation of dissolved Mn(II) has been practically applied to the removal of Mn(II) from groundwater. Some water purification plants have sand filter facilities wherein microbially mediated precipitation of Fe and Mn occurs.[4] The removal of Mn(II) is conventionally achieved by inorganic oxidation such as chlorination or permanganate oxidation, followed by sand filtration. In recent years, research has illustrated the advantage of biological Fe and Mn removal processes for groundwater treatment. It can minimize the addition of chemical reagents, resulting in the reduction of operational costs and also by-products formation.[4,5] Recent investigation on the interaction of trace metals with biogenic Mn oxides has raised important implications for the remediation of waters, soils, and sediments contaminated with metals.[3,6,7] Biogenic Mn oxides sequester numerous metal ions (e.g., Ni, Zn, Cu, Co, Pb, Cd, As, Cr, and U). Several attempts have been made to develop bioreactors that retain microbial Mn(II)-oxidizing consortia. Such studies gained interest in the biotechnological application of microbial Mn(II) oxidation to treatment of waters contaminated with Mn(II) and other metals.

1.2 Microbial Mn(II) Oxidation

In bacteria and fungi the Mn(II) oxidation occurs extracellularly or at the outermost layers of cells, resulting in the precipitation of Mn oxides on cell surfaces (Fig. 1.1). The multi-copper containing oxidases (MCOs) are a class of Mn(II) oxidases. Increasing evidence indicates the widespread distribution of MCO-type Mn(II) oxidases over bacterial Mn oxidizers (i.e., MnxG, MofA, CumA, and MoxA from *Bacillus* sp. SG-1, *Leptothrix discophora*, *Pseudomonas putida*, and *Pedomicrobium* sp. ACM3067, respectively).[2,8] Several other proteins and small molecules also are involved.[2] The role of peroxidase, an iron (heme) containing protein, has been recently found in the Mn(II) oxidation by marine bacteria, *Aurantimonas manganoxydans* and *Erythrobacter* sp. SD-21.[9]

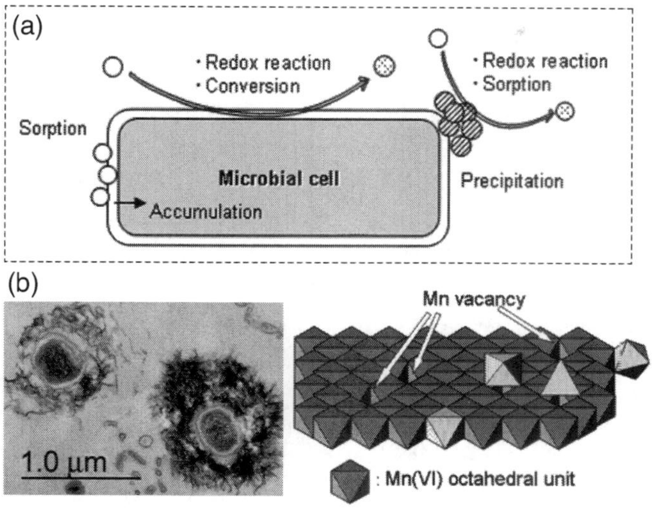

Figure 1.1. (a) Different modes of microbe–metal interactions, including sorption, accumulation, and conversion and precipitation by microbially mediated reactions. The precipitates (e.g., biogenic Fe and Mn oxide minerals) also serve as reactive agents such as adsorbents, oxidants or reductants, and catalysts. (b) Precipitation of biogenic Mn(IV) oxides in outermost layers of bacterial cells. The biogenic oxides (shown as filaments in the transmission electron micrograph) have nanocrystalline, negatively charged layer structures, leading to high affinity and capacity for sorption of metal cations such as Zn(II), Co(II), Ni(II), and Pb(II) (see the text).

A similarity between enzymatic Mn(II) oxidation pathways in bacteria and fungi has been argued.[1] In practice recent research showed that the MCO-type oxidases play the roll in phylogenetically diverse ascomycete fungi, including *Acremonium* sp. KR21-2.[10,11] The fungal Mn oxidase has been purified to homogeneity and used for elucidating the Mn(II) oxidation process.[12]

While microbial Mn(II) oxidation is widespread, little is known about why microorganisms oxidize Mn(II). There has been no conclusive evidence for autotrophic growth coupling to Mn(II) oxidation in bacteria. Recent research has strongly suggested the energetic benefit from Mn(II) oxidation in the *A. manganoxydans*.[13]

In bacterial species *Bacillus* sp. SG-1, *P. putida*, and *L. discophora*, X-ray absorption spectroscopy and other analytical techniques have established the detailed structural models for biogenic Mn oxides. These models represent nanocrystallized birnessite or vernadite (δ-MnO_2).[1,14–16] Birnessite has a structural unit of sheet of MnO_6 octahedra with a 7 Å interlayer spacing[17] (Fig. 1.1). The interlayer spacings are occupied by charge-balancing cations such as H^+, Na^+, K^+, Mn^{2+}, and Ca^{2+} and water molecules. δ-MnO_2 has the same structural unit, but is disordered in the layer stacking direction or is composed of extremely thin plates.[17] In fungi, *Acremonium* sp. KR21-2 and other aquatic ascomycetes appear to form δ-MnO_2 in liquid cultures. However, strain KR21-2 was recently reported to produce a todorokite-like mineral that has tunnel structures at different culture conditions.[18,19] Culture conditions may affect the Mn oxide structure.

1.3 Interaction of Biogenic Mn Oxides with Trace Metal Ions

Biogenic Mn oxides exhibit high affinity and capacity for sorption of metal cations such as Zn(II), Co(II), Ni(II), and Pb(II),[20–22] owing to their high specific surface area (100–250 m^2/g) and the negatively charged layer structures. The Pb(II) sorption capacity of *L. discophora* oxide was found to be Pb/Mn molar ratio of up to 0.5,[20] which is several orders of magnitude higher than that of commercially available MnO_2. A high specific surface area of *L. discophora* oxide (224 m^2/g), compared with those of commercially available oxides (<4.7 m^2/g), is responsible for the large Pb(II) sorption capacity. In addition, the Mn oxide's structural unit has an abundance of Mn(IV)-vacancies (Fig. 1.1), leading to the occurrence of high negative-charge density. Such vacancies in *P. putida*[15] and *L. discophora*[16] oxides have been estimated to reach up to 12–17% for Mn(IV) structural unit. A study with cultures of *Acremonium* sp. KR21-2 demonstrated the high sorption ability of biogenic Mn oxides for Zn(II), Ni(II), and Co(II).[21] For example, the molar ratios of sorbed Zn and Co to solid Mn reached 0.3 and 0.23, respectively. The high sorption ability for Zn(II) and Ni(II) was also reported in a deep-sea sedimentary bacterium.[23] The sorbed Co likely exists as Co(III) species because Co(II) is oxidized readily by biogenic Mn oxides.[24] A very high sorption capacity for Cd(II) (Cd/Mn molar ratio up to 0.69) was demonstrated using the biogenic Mn oxide of a *Bacillus* strain.[25]

Contamination of groundwater with As is one of the important environmental problems worldwide. Iron hydroxides are known to adsorb As(V) but is less effective for the more toxic As(III). Biogenic Mn oxides act as a powerful oxidant for As(III),[26] so that the presence of Mn oxides could improve the treatment of As-contaminated groundwater. Tani *et al.* demonstrated that in liquid cultures of *Acremonium* sp. KR21-2, the biogenic Mn oxide formation resulted in the As(III) oxidation to As(V).[26] Coexisting cations such as Mn(II) and Zn(II) largely affected the oxidation rate of As(III) to As(V), as well as the sorption efficiency for As(V).

Naturally occurring and chemically synthesized Mn oxides have been shown to oxidize other elements, including Ce(III) to Ce(IV),[27] Sb(III) to Sb(V),[28] and Se(IV) to Se(VI).[29] Thus, Mn-oxidizing microorganisms can participate indirectly in the fates of these elements via the Mn oxide formation.[24,26] In addition, several elements that commonly exist as oxyanion forms, such as W, Mo, V, Sb, and Te, are known to accumulate in natural metal oxide phases.[30–32] Detailed studies on the interactions of biogenic Mn oxides with these elements will lead to new applications.

Biogenic Mn oxide production may be applicable to removal of radionuclides from waters. It has been demonstrated that the Mn oxide of *Bacillus* sp. SG-1 incorporates uranyl ion [$U(VI)O_2^{2+}$] up to a U/Mn molar ratio of 0.32.[33] Interestingly, the structure of biogenic oxide changes from the sheet- to a tunnel-type to form more stable U(VI) complex on Mn oxide.

1.4 Investigations with Mn(II)-Oxidizing Bioreactors

Application of microbial Mn(II) oxidation to remediation of contaminated waters has been examined with bioreactors that retain microbial Mn(II)-oxidizing consortia under nonsterilized conditions. Such consortia have been obtained from freshwater Mn precipitates (e.g., Mn oxide coatings on pebble surfaces,[34–36] which are good sources for culturing Mn oxidizers). However, a freshwater epilithic biofilm that did not contain Mn oxide also readily yielded a Mn-oxidizing consortium[34] likely because of widespread distribution of Mn oxidizers in environments. Bioreactor operations result in enriching Mn oxidizers, so that the Mn(II) oxidation rates increase gradually.[34,36] Mariner *et al.* conducted the bioreactor experiment examined over one year and showed that it was effective in removing Mn(II) from 10 to <0.25 mg/L.[36] In addition, the reactor could function over a wide range of temperatures from 5°C to 30°C and a pH range from 4 to 6. The authors suggested the use of the Mn(II) oxidation process for treatment of Mn-contaminated waters and acid mine drainage waters. The consortia contained various Mn oxidizers: ascomycete fungi and a bacterium phylogenetically close to *Bosea thiooxidans* have been isolated.[36]

Miyata *et al.* characterized the biogenic oxides produced in freshwater Mn(II)-oxidizing consortia maintained over one year. The Mn valences in Mn oxides were close to +4, and the specific surface area was $\sim$120 m^2/g.[34] The repeated batch cultures of consortium resulted in the simultaneous immobilization of dissolved Mn(II) and a second metal cation, Zn(II) or Ni(II) added at a low concentration (0.1 mg/L) (Fig. 1.2). In cultures fresh Mn oxides are supplied continuously by microbial activity, and these can serve as an effective adsorbent for the metal cations. It should be noted that the sorption of trace metal ions was not hindered by the presence of approximately 50-fold excess of dissolved Mn(II) (Fig. 1.2). Thus, such concentrations of Zn(II) and Ni(II) (and also Co(II); N. Miyata and Y. Tani, unpublished data) do not compete with excess Mn(II) for the adsorption sites on biogenic Mn oxides. This is favorable for recovery of metals present only at trace levels. However, addition of Zn(II) or Ni(II) at higher concentrations (1 mg/L) inhibited the Mn(II) oxidation activity in consortium, so that the removal efficiencies of dissolved Zn(II) and Ni(II) decreased gradually.[34] The inhibitory effect of higher concentrations of metals on the microbial Mn(II) oxidation activity remains to be resolved.

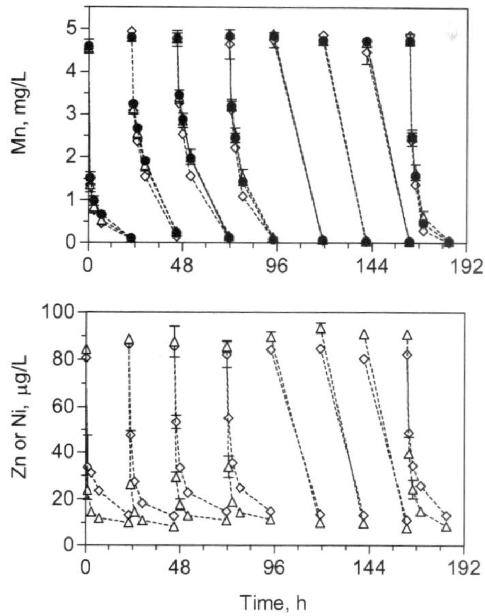

Figure 1.2. Removal of dissolved Mn(II) and Zn(II) or Ni(II) by repeated batch culture of microbial consortium enriched from freshwater epilithic biofilm. Open triangles and diamonds represent the remaining concentrations of dissolved metal ions in the cultures with Zn(II) and Ni(II), respectively. Closed circles represent the remaining concentration of dissolved Mn(II) in the control culture (without Zn or Ni). Adapted from previous report.[34]

1.5 Conclusions and Outlook

The Mn(II)-oxidizing microorganisms sequester numerous metals and other elements indirectly via the production of highly reactive agents capable of serving as adsorbents, oxidants, and catalysts. The microbial Mn(II) oxidation could be a promising tool for recovery of metals and remediation of metal-contaminated waters. Further research is required to clarify the toxic effects of metals on the microorganisms and their activity of Mn(II) oxidation. Many interesting elements including rare earth elements and radionuclides as well as toxic metal(loid)s are potent targets in the use of microbial Mn(II) oxidation. Furthermore, the biogenic Mn oxide production has been examined for oxidation of small organic pollutants.[37]

Given that diverse microorganisms oxidize Mn(II) at different environmental conditions and also different enzymes (MCO- and peroxidase-types) participate in the oxidation, it is likely that structural diversity of biogenic minerals with inherent properties are offered for various applications. Furthermore, biogenic Mn oxide phases synthesized in microbial cultures may be collected and used as inorganic materials for innovative technologies.[7,19] For this purpose, separation of pure precipitates from the microbial cellular components would be an important subject. Synthesis of Mn(IV) oxide using purified (and also cloned) enzyme rather than microbial cell may be favorable; such investigation has been considered very little as yet. Further research on the microbial Mn(II) oxidation from the biotechnological viewpoint opens a window for new applications of microbe–metal interactions.

Acknowledgments

We acknowledge support by Grant-in-Aid for Scientific Research (C) from the Japan Society for the Promotion of Science (No. 18510080 and 19510091).

References

1. Tebo, B. M., Bargar, J. R., Clement, B. G., Dick, G. J., Murray, K. J., Parker, D., Verity, R., and Webb, S. M., *Annu. Rev. Earth Planet. Sci.* **32**, 287 (2004).
2. Tebo, B. M., Johnson, H. A., McCarthy, J. K., and Templeton A. S., *Trends Microbiol.* **13**, 421 (2005).
3. Miyata, N., Tani, Y., Sakata M., and Iwahori, K., *J. Biosci. Bioeng.* **104**, 1 (2007).
4. Mouchet, P., *J. Am. Water Works Assoc.* **84**, 158 (1992).
5. Katsoyiannis I. A. and Zouboulis, A. I., *Water Res.* **38**, 1922 (2004).
6. Villalobos, M., Bargar J., and Sposito, G., *Elements* **1**, 223 (2005).
7. Hennebel, T., de Gusseme, B., Boon, N., and Verstraete, W., *Trends Biotechnol.* **27**, 90 (2009).
8. Ridge, J. P., Lin, M., Larsen, E. I., Fegan, M., McEwan A. G., and Sly, L. I., *Environ. Microbiol.* **9**, 944 (2007).

9. Anderson, C. R., Johnson, H. A., Caputo, N., Davis, R. E., Torpey J. A., and Tebo, B. M., *Appl. Environ. Microbiol.* **75**, 4130 (2009).
10. Miyata, N., Maruo, K., Tani, Y., Tsuno, H., Seyama, H., Soma M., and Iwahori, K., *Geomicrobiol. J.* **23**, 63 (2006).
11. Miyata, N., Tani, Y., Iwahori, K., and Soma, M. *FEMS Microbiol. Ecol.* **47**, 101 (2004).
12. Miyata, N., Tani, Y., Maruo, K., Tsuno, H., Sakata, M., and Iwahori, K. *Appl. Environ. Microbiol.* **72**, 6467 (2006).
13. Dick, G. J., Podell, S., Johnson, H. A., Rivera-Espinoza, Y., Bernier-Latmani, R., McCarthy, J. K., Torpey, J. W., Clement, B. G., Gaasterland T., and Tebo, B. M., *Appl. Environ. Microbiol.* **74**, 2658 (2008).
14. Bargar, J. R., Tebo, B. M., Bergmann, U., Webb, S. M., Glatzel, P., Chiu V. Q., and Villalobos, M., *Am. Mineral.* **90**, 143 (2005).
15. Villalobos, M., Lanson, B., Manceau, A., Toner, B., and Sposito, G., *Am. Mineral.* **91**, 489 (2006).
16. Saratovsky, I., Wightman, P. G., Pastén, P. A., Gaillard J.-F., and Poeppelmeier, K. R., *J. Am. Chem. Soc.* **128**, 11188 (2006).
17. Post, J. E., *Proc. Natl. Acad. Sci. USA* **96**, 3447 (1999).
18. Saratovsky, I., Gurr, S. J., and Hayward, M. A., *Geochim. Cosmochim. Acta* **73**, 3291 (2009).
19. Petkov, V., Ren, Y., Saratovsky, I., Pastén, P., Gurr, S. J., Hayward, M. A., Poeppelmeier K. R., and Gaillard, J.-F., *ACS Nano* **3**, 441 (2009).
20. Nelson, Y. M., Lion, L. W., Ghiorse W. C., and Shuler, M. L., *Appl. Environ. Microbiol.* **65**, 175 (1999).
21. Tani, Y., Ohashi, M., Miyata, N., Seyama, H., Iwahori K., and Soma, M., *J. Environ. Sci. Health A* **39**, 2641 (2004).
22. Villalobos, M., Bargar J., and Sposito, G., *Environ. Sci. Technol.* **39**, 569 (2005).
23. Wang, W., Shao, Z., Liu, Y., and Wang, G., *Microbiology* **155**, 1989 (2009).
24. Murray, K. J., Webb, S. M., Bargar, J. R., and Tebo, B. M., *Appl. Environ. Microbiol.* **73**, 6905 (2007).
25. Meng, Y. T., Zheng, Y. M., Zhang, L. M., and He, J. Z. *Environ. Pollut.* **157**, 2577 (2009).
26. Tani, Y., Miyata, N., Ohashi, M., Ohnuki, T., Seyama, H., Iwahori K., and Soma, M. *Environ. Sci. Technol.* **38**, 6618 (2004).
27. Takahashi, Y., Shimizu, H., Usui, A., Kagi H., and Nomura, M., *Geochim. Cosmochim. Acta* **64**, 2929 (2000).

28. Belzile, N., Chen, Y. W., and Wang, Z. *Chem. Geol.* **174**, 379 (2001).
29. Scott, M. J., and Morgan, J. J., *Environ. Sci. Technol.* **30**, 1990 (1996).
30. Müller, B., Granina, L., Schaller, T., Ulrich, A., and Wehrli, B., *Environ. Sci. Technol.* **36**, 411 (2002).
31. Tani, Y., Miyata, N., Iwahori, K., Soma, M., Tokuda, S., Seyama, H., and Theng, B. K. G., *Appl. Geochem.* **18**, 1541 (2003).
32. Kuhn, T., Bostick, B. C., Koschinsky, A., Halbach P., and Fendorf, S., *Chem. Geol.* **199**, 29 (2003).
33. Webb, S. M., Fuller, C. C., Tebo B. M., and Bargar, J. R., *Environ. Sci. Technol.* **40**, 771 (2006).
34. Miyata, N., Sugiyama, D., Tani, Y., Tsuno, H., Seyama, H., Sakata, M., and Iwahori, K., *J. Biosci. Bioeng.* **103**, 432 (2007).
35. Hallberg, K. B. and Johnson, D. B., *Sci. Total Environ.* **338**, 115 (2005).
36. Mariner, R., Johnson, D. B., and Hallberg, K. B., *Hydrometallurgy* **94**, 100 (2008).
37. de Rudder, J., van de Wiele, T., Dhooge, W., Comhaire, F., and Verstraete, W., *Water Res.* **38**, 184 (2004).

Chapter 2

BIOLOGICAL TREATMENT FOR REMOVAL OF IRON, MANGANESE, AND ARSENIC FROM GROUNDWATER

Ichiro Suzuki and Danladi Mahuta Sahabi

Department of Chemical and Energy Engineering, Graduate School of Engineering, Yokohama National University, 79-5 Tokiwa-dai, Hodogaya, Yokohama, Kanagawa 240-8501, Japan

suz-1@ynu.ac.jp

2.1 Introduction

Many groundwater supplies are contaminated by varying levels of iron (Fe), manganese (Mn), and arsenic (As) in concentrations that exceed the drinking water guidelines. Fe and Mn are removed from water supplies mainly for aesthetic and operational objectives, whereas As must be removed from drinking water because of its numerous health hazards. Conventional treatment methods for removal of Fe^{2+} and Mn^{2+} from groundwater are based on physicochemical processes such as aeration, chemical oxidation, coagulation, and rapid sand filtration. An alternative method is based on biological process of Fe and Mn oxidation and deposition mediated

Handbook of Metal Biotechnology: Applications for Environmental Conservation and Sustainability

Edited by Michihiko Ike, Mitsuo Yamashita, and Satoshi Soda

www.panstanford.com

by microbial consortia fixed in biological filtration reactor. In addition, this method is possibly applicable to As removal, since soluble As oxyanions are known to be adsorbed onto Fe and Mn oxides produced by the biological treatment process. In this chapter, research advances toward unraveling the mechanisms of biological treatment of Fe and Mn, as well as the potential application of biological Fe and Mn oxidation processes for As removal from groundwater are discussed.

2.2 Biological Removal of Fe and Mn

Fe and Mn usually exist in groundwater as soluble Fe^{2+} and Mn^{2+}, and their removal is based on their oxidation into Fe(III) and Mn(IV) states, which are characterized by low solubility and therefore easily removed by sedimentation and filtration. Conventional Fe and Mn removal plants rely on physical-chemical reactions using Mn-greensands, intense aeration, or chemical oxidation (with O_3, $KMnO_4$, or ClO_2). An alternative process is based on biological method using Fe- and Mn-oxidizing microorganisms.[1] Advantages of biological treatment over the conventional physical-chemical methods include less use of chemicals, reduction of excess sludge, and lower operation and maintenance costs.[1,2]

2.2.1 *Biological Treatment Plant for Removal of Fe and Mn*

For many years, improved Fe and Mn removal was observed in traditional slow sand filtration systems and several conventional treatment plants when the so-called iron bacteria were grown in the filters.[1,3,4] Tanimoto described that removal of Fe and Mn was observed when iron bacteria, such as Fe-oxidizing *Gallionella* sp. or Fe- and Mn-oxidizing *Leptothrix* sp., were grown in slow sand filtration system at Tadotsu-cho, Kagawa, in Japan (established in 1924), which was, perhaps, the first finding on biological Fe and Mn filtration.[3,4] Consequently, inoculation of slow sand filters with iron bacteria was practiced to improve the removal of these metals at several groundwater treatment systems in Japan.[3] During same period, the removal of Fe and Mn by microbial activity, using slow sand filtration without addition of chemical oxidants, was also known in several European waterworks.[1,5] The first modern biological treatment systems using rapid sand filtration for treatment of Fe and Mn from groundwater was probably developed in France in the 1980s.[1] Nowadays biological processes to remove Fe and Mn are widely used in Europe, United States, Canada, and Japan and there are some treatment plants in China, Africa, and Latin America.[1,2,5–13]

In Europe most biological filtration plants employ two filtration steps to remove Fe and Mn when both metals are present in the groundwater, because of the differences in optimum conditions required for biological Fe and Mn oxidation. The steps include initial aeration followed by rapid filtration for Fe removal and secondary aeration, pH adjustment, and secondary rapid filtration for Mn removal.[1] These systems require sophisticated devices to control dissolve oxygen concentration, pH, and redox potential of each filtration steps. In Japan, however, single filtration systems are commonly used in biological treatment plants for simultaneous removal of Fe and Mn. The first of such model biological treatment plant was constructed in 1993, in Joyo city of southern Kyoto prefecture. The system consisted of aeration by dropping, gravity filtration with dual-media filter (anthracite and sand), and final disinfection. The filtration rate of the reactor is 2.9 m/h.[2] Another biological Fe and Mn treatment system with rapid filtration rate is located in Yamato-Koriyama city, Nara (Established in 2001). The system is consisted of aeration by dropping and gravity filtration with single filter using spherical carriers of polyester fiber, which enable the filtration rate of up to 15 m/h.[8] These single filter systems are easy to operate, but may require special filter media.

2.3 Principles of Biological Fe and Mn Removal

The terms biological and physicochemical processes are not mutually exclusive in groundwater treatment for removal of Fe and Mn. In biological water treatment method, soluble Fe^{2+} and Mn^{2+} are removed through oxidation and filtration of the insoluble metal oxides. At neutral pH, Fe^{2+} is easily oxidized by dissolved oxygen, thus Fe^{2+}

oxidation is mainly achieved through aeration; in contrast, Mn^{2+} oxidation does not proceed at appreciable rates below pH 9.[1] Under such conditions, biological process should be responsible for Mn oxidation.[1,14] In addition to direct microbial oxidation, substantial amounts of Mn may be removed by the pre-formed biogenic Mn oxides through adsorption and autocatalytic oxidation.[10,15,16] Thus, removal of Fe and Mn by biological treatment is achieved through a combination of physicochemical and biological removal processes, which include chemical and biological oxidation, biosorption and adsorption, as well as autocatalytic oxidation.[1,10,15,16]

2.3.1 *Maturation and "Aging" of Biological Filter Media*

The major distinguishing feature of biological phenomenon from the physical-chemical one is the requirement of the former for maturation - a period for the formation of oxides-rich biofilm on the filter media. Once a biological Fe or Mn removal plant is constructed, the system must be given time to mature. The coating of the media surface with microbial consortia and catalytic oxides layer is a spontaneous process that depends on the chemical and biological characteristics of the groundwater.[1,2,11] For Fe removal plants, the maturing period is quite short requiring anywhere from a day to about one week. In well-aerated systems, complete removal of Fe sometimes occurs before the formation of the microbial layer, indicating that substantial Fe oxidation is achieved through aeration.[1,17] For Mn removal plants the time can be considerably longer, anywhere from 2 weeks to 3 months.[1,2,11] Biological activity may be essential for initiation of Mn oxidation in newly constructed biological treatment plants. After Mn oxidation is initiated by the microbial activity, part of the biogenic Mn oxides (and Fe oxides, if present) is continuously deposited as oxides(catalytic) layers coating the filter media. Biofilter media coated with significant amounts of metal oxides are termed as "aged" and can efficiently oxidize Mn without the chemical oxidants.[16]

Table 2.1 shows the surface coatings compositions of two groups of aged biofilter media collected from Joyo biological treatment plant.[16,18] The plant was established in 1993 with three filter wells and later extended with additional three wells in 2005. At the time

Table 2.1. Compositions of surface coatings on Joyo biological filter media.[16,18]

Measured parameters	3-year filter media	15-years filter media
Mass of surface coating (mg/g)	31.7 ± 2.9	78.3 ± 10.4
Fe (mg/g)	3.4 ± 0.5	12.9 ± 0.6
Mn (mg/g)	5.0 ± 0.3	19.9 ± 3.3
Total proteins (mg/g)	0.556 ± 0.219	0.545 ± 0.068
Total 16S rRNA genes ($\times 10^8$ copies/g)	6.09 ± 1.01	6.02 ± 1.86

of sampling (January 2008), the first group of the biofilter media has been in use for about 15 years, while the second group for 3 years. As seen in Table 2.1, continuous deposition of Fe and Mn oxides on the surface of the filter media, despite regular backwashing, resulted to about fourfold increase in concentrations of Fe and Mn oxides (combined) between the 3- and 15-year filter media. Unlike the catalytic oxides layer, the mean values of total proteins and copy numbers of 16S ribosomal RNA (rRNA) genes as estimated by real-time polymerase chain reaction (PCR) were quite similar per gram of both the 3- and 15-year biofilter media (Table 2.1). This may mean that the growth of the bacterial community on the surface of the biofilter media is regulated by the operational conditions of the plant and that the biofilms has to renew little by little after regular backwashing.

2.3.2 *Physicochemical and Biological Removal of Mn by Aged Biofilter*

The roles of physicochemical processes in Mn removal may be a function of the accumulated oxides layer, which, under similar conditions as in the case of the two groups of biofilter media in Joyo, can be attributed to the operation time of the filter media. The amounts of soluble Mn^{2+} removed over 24 hours by 1 g each of the 3- and 15-year filter media were 0.34 and 1.17 mg, respectively.[16] However, it should be noted that surface coatings compositions of biofilter media are heterogeneous; it is therefore impossible to compare the performance of these complex materials simply based on duration of operation, more so if the biofilters are from different sources. For example, in contrast to Joyo biofilter media, Vandenabeele *et al.*[12]

concluded that abiotic physicochemical processes were not sufficient to account for the observed rates of Mn removal by the biofilter media of a biological treatment plant in Snellegem, Belgium.

Obviously, the catalytic Mn oxide coatings on the surface of biofilter media should be produced by microbial activities at the start up of the biological reactors. In addition, biological re-oxidation of Mn is required to replenish the adsorption and oxidation powers of the catalytic oxide layers. Catalytic oxidation of Mn^{2+} by Mn oxides causes reduction of tetravalent Mn oxides to lower oxidation states, and if not re-oxidized, the Mn oxides can lose their adsorption and autocatalytic activity with time. Similar loss of oxidation activity is seen in conventional Mn removal systems using contact oxidation by Mn greensand, and chemical oxidants such as potassium permanganate or chlorine are required to reactivate the oxides layer of filter media. In a way not yet clearly understood, microbial activity should be responsible for maintaining the adsorptive and autocatalytic powers of oxides-coated biofilter media. This concerted action of microbial activity and physicochemical processes make it possible to recover Mn removal activity within one hour after backwashing of the filter bed[2,8] and could possibly explain the robustness of biological filtration systems at different seasons of the year.

2.4 Microorganisms in Biological Removal of Fe and Mn

The Fe- and Mn-oxidizing microbial community is the core of the technology for the biological removal of Fe and Mn from ground water. The commonly called iron bacteria, such as the genera *Gallionella*, *Leptothrix*, *Sphaerotilus*, *Crenothrix*, *Clonothrix*, *Siderocapsa*, and *Metallogenium*, are thought to be involved in biological Fe and Mn removal; however, their characteristics of metal oxidation or metal sorption activities and contribution to the removal of Fe and Mn in biological treatment plants is still unclear. These iron bacteria have been well known by their unique structures such as twisted filaments (*Gallionella* and *Toxothrix*), filamentous tubes called "sheath" (*Leptothrix*, *Sphaerotilus*, *Clonothrix*, and *Crenothrix*), and star-shaped structure with Mn oxide minerals (*Metallogenium*). But their characteristics of growth conditions and biological activities are still uncertain because of the difficulties of their cultivations except the genera *Gallionella*, *Leptothrix*, and *Sphaerotilus*. In addition to the Fe- and Mn-oxidizing microorganisms, the activity of nitrifying and denitrifying bacteria in the biofilm may be critical, at least for initiation of Mn oxidation.[1,12,17,19,20]

2.4.1 *Iron Bacteria*

Iron bacteria have long been well known not only for their interesting habitats of Fe- and Mn-rich deposits and environments but also for their unique shapes. *Gallionella ferruginea* is an Fe-oxidizing chemolithoautotroph which has unique spirally twisted stalks.[21] *G. ferruginea* is able to utilize Fe^{2+} oxidation for its growth, whereas it dose not oxidize Mn^{2+}. *Leptothrix* sp. are Fe- and Mn-oxidizing, sheathed heterotrophs.[14,22] *Leptothrix* sp. are able to oxidize both Fe^{2+} and Mn^{2+}; however, they can not use these metals as energy sources. A number of *Leptothrix* species are well investigated in their characteristics by success of their pure cultures; however, pure culture of *Leptothrix ochracea*, which is the most common species found in biological treatment reactors, has not been obtained yet.[14,22] The genus *Sphaerotilus* is phylogenetically related to the genus *Leptothrix*; however, it does not oxidize Mn^{2+}.[14,22] *Gallionella*, *Leptothrix*, and *Sphaerotilus* belong to beta subclass of Proteobacteria according to phylogenetic analyses of 16S rRNA gene sequences. The other Mn-oxidizing bacteria found in biological treatment reactors are members of the genus *Hyphomicrobium*, which are budding (prosthecate) bacteria belonging to alpha subclass of Proteobacteria.[23–25] The other iron bacteria, such as *Clonothrix*,[26] *Crenothrix*,[27] *Toxothrix*,[28,29] *Shiderocapsa*,[30] and *Metallogenium*,[31,32] generally exist in Fe and Mn oxide deposits; however, their metal oxidation activities and contribution for removal of Fe and Mn in biological treatment of groundwater are unclear. *Toxothrix* and *Siderocapsa* are not yet isolated in pure culture, and *Metallogenium* is recently considered as Mn oxide particle produced by the other Mn-oxidizing microorganisms rather than a bacterium.[31,32]

Such iron bacteria are easily found in microbial deposits on the surface of the biological treatment reactors by microscopic observations;[1,2,4,5,11] however, viability of these bacterial cells were not reported yet. Probably, most of the cells in the deposits are just accumulated on the filter through raw groundwater but do not grow on the filter. As described above, Fe and Mn removal activities of the biofilters are recovered within one hour after backwashing,[2,8] suggesting that the microorganisms mainly contribute to Fe and Mn removal are existing in biofilms developed on the surface of the biofilter rather than the cells existing in the deposits.

2.4.2 *Nitrifying and Denitrifying Microorganisms*

It is known that initiation of biological Mn oxidation requires complete nitrification-denitrification, especially if the water contains relatively high concentration of ammonia.[1,12,17,19,20,33] Gouzinis *et al.*[33] investigated the biological oxidation of Fe, Mn, and ammonium when present simultaneously in water. They found that high concentrations of Fe and ammonia affect Mn oxidation negatively, and Mn removal occurs after complete nitrification has taken place. An indirect link between nitrification-denitrification and Mn oxidation was reported by Vandenabeele *et al.*,[12] that presence of nitrate increased Mn-oxidizing activities in batch cultures of microbial consortia derived from biological filter media. However, how nitrate influence the activity of Mn-oxidizing bacteria in biological treatment system remains unknown.

2.4.3 *Molecular Diversity of Bacterial Consortia Involved in Biological Removal of Fe and Mn*

The development of analytical methods for studying microbial communities based on culture-independent molecular biological techniques has further revealed the presence of microorganisms in biological water treatment plants that were not hitherto detectable by classical microscopic and cultivation techniques, thus changing the previous concepts on the structures and function of microbial communities involved in Fe and Mn removal. The microbial diversity in Joyo biological treatment plant was investigated using PCR-based 16S rRNA gene analyses such as clone library analyses and real-time PCR.[18,34] Total microbial DNA was extracted from the biofilm coated

Table 2.2. List of bacteria detected by phylogenetic analysis of 16S rRNA gene clone library from Joyo biological treatment plant.[18 34]

	Closest relatives of the clones
Fe- and Mn-oxidizing bacteria	*Leptothrix mobilis, Caldimonas manganoxidans, Hyphomicrobium* sp.
Fe-oxidizing bacteria	*Gallionella ferruginea, Thiobacillus denitrificans*
Fe-reducing bacterium	*Geothrix fermentans*
Ammonia-oxidizing bacteria	*Nitrosomonas* sp., *Nitrosospira* sp.
Nitrite-oxidizing bacterium	*Nitrospira* sp.
Denitrifying bacteria	*T. denitrificans, Azoarcus denitrificans*

on the biofilter media, and a clone library of bacterial 16S rRNA genes was constructed. DNA sequence analyses of the individual clones (Table 2.2) revealed that bacteria closely related to Fe-oxidizing bacterium *Gallionella* and Fe- and Mn-oxidizing bacteria *Leptothrix, Hyphomicrobium, Caldimonas*[35] were observed. Also detected were denitrifying bacteria *Thiobacillus denitrificans* which perform nitrate-dependent Fe oxidation,[36] ammonia-oxidizing bacteria *Nitrosomonas* and *Nitrosospira*, nitrite-oxidizing bacterium *Nitrospira*, denitrifying bacteria *Azoarcus denitrificans*, and Fe-reducing bacterium *Geothrix fermentans*. Although the functional roles of each of these bacteria could not be ascertained from their presence as revealed by the culture-independent techniques, the necessity of symbiotic interdependence between the different autotrophs and heterotrophs is foreseeable as a survival strategy in the carbon- and nitrogen-deficient groundwater environment. Further analyses to estimate the amount of the bacteria closely related to *Leptothrix* on Joyo biofilter media were performed by real-time PCR using oligonucleotide probe PSP-6[37] as one of the primers, which is specific for 16S rRNA genes of *Leptothrix* sp. The result showed that the rates of 16S rRNA genes of *Leptothrix*-related bacteria to those of total bacteria in Joyo biofilter media ranged between 3 and 11.3%.[18] Similar real-time PCR analysis using the PSP-6 primer was carried out by Burger *et al.*[7] to estimate the amount of *Leptothrix*-related bacteria in biofilter media collected from four different biological treatment plants in New Brunswick, Canada. Interestingly, the amplification of 16S rRNA genes of *Leptothrix*-related bacteria was observed only in one out of their four samples.[7] The implication of these data is that *Leptothrix* sp. could be absent or, at least a minority, in many biological Fe and Mn removal plants. Obviously, detail evaluation of the bacteria responsible for Fe and Mn removal in biological treatment plants awaits further investigations.

2.5 Integration of Biological Fe and Mn Oxidation with As Removal

Contamination of groundwater with As is one of the most important environmental problems worldwide. In Japan, 2.1% of the 3663 wells sampled had As concentrations above the World Health Organization (WHO) guideline value (10 μg/L).[38] In natural water As is mainly present in inorganic forms of oxianions as arsenate [As(V)] and arsenite [As(III)] species. Arsenates are the major species in oxygenated waters, whereas arsenites predominate under reduced conditions such as found in groundwaters. As(III) is more toxic to biological systems and more mobile in soil and aquatic environments than As(V). The conventional techniques for As removal from water such as precipitation-coagulation, membrane separation, and ion-exchange are generally expensive and less effective for As(III). Therefore, pre-oxidation of As(III) to As(V) is often required for effective removal of As(III) from water. Since most As contaminations in groundwaters are permanent problems induced by natural mineral sources or mining activities, low-cost and easy-to-operate removal systems are required, especially for developing economies. Biological sorptive filtration, which integrates biological Fe and Mn oxidation with As removal from groundwater, combines these qualities and may be the method of choice for remediation of this natural calamity. The method comprises of the steps of biological oxidation of Fe and Mn forming amorphous Fe and Mn oxides, which coat the surface of the filter media.[39,40] Not only As(V) but also As(III) are removed from the groundwater by a combination of biological and physicochemical sorption processes, including oxidation and adsorption onto the biogenic Fe and Mn oxides.[39–41] The biological approach is more economical and eco-friendly as no chemicals are added and the sorbent materials are continuously produced *in situ*. Furthermore, it is a combined process for simultaneous removal of Fe, Mn, and As from groundwater.[20,42] High concentrations of As in groundwaters are often associated with high Fe and Mn levels; therefore, biological sorptive filtration system using biogenic Fe and Mn oxides is suitable for simultaneous removal of As, Fe, and Mn.

2.5.1 *Removal of As by Biological Sorptive Filtration*

The physicochemical aspects of biological sorptive filtration, including abiotic oxidation of As(III) and adsorption of both As(III) and As(V), were well investigated using the oxides-coated biofilter media collected from Joyo.[40] The biofilter media can easily oxidize As(III) to As(V), and the rate of oxidation is less affected by pH variations from 4 to 8.5; however, the retention capacity of the media for the produced or added As(V) depends strongly on the pH of the solution. The high rates of As(III) oxidation were attributed to the high Mn/Fe ratio of the biofilter media, whereas the effects of pH on the adsorption of arsenates was explained on the basis of the point of zero charge of the medium.[40] Depending on the experimental conditions, the biologically coated media can remove As from water through a wide range of mechanisms, including specific adsorptions, ligand exchange, and electrostatic attractions.

2.5.2 *Pilot Plant Studies for Simultaneous Removal of Fe, Mn, and As by Biological Processes*

A number of active studies on simultaneous removal of Fe, Mn, and As, mostly at the levels of pilot plants, have been reported across the world.[20,42–44] These studies have shown that As concentration in the groundwaters were successfully decreased to the level under the WHO's standard by the biological processes. One of such pilot plants which has been maintained in "M-city" in Japan is operated at a filtration rate of 25 m/h using polypropylene tubes (ca. 5 mm) as filter media.[20] The plant achieves the removal of Fe and Mn with concurrent reduction of As concentration from 22.5 to 5.5 μg/L. This pilot plant consists of simple reactors which are easy to operate. However, as described above, mechanism of Fe and Mn oxidation by microbial consortia in biological treatment system are still unknown. Further investigations are therefore needed to fully understand the mechanisms of these complex systems with the view to improve the efficiency of this low-cost As removal technology and possible extends to a lot of other toxic contaminants from water.

2.6 Conclusion

Biological oxidation is becoming a conventional treatment method for removal of Fe and Mn from groundwater in many countries around the world, because of its numerous advantages.

- The method combines the oxidative potentials of native Fe- and Mn-oxidizing microbial consortia in groundwater with the physicochemical oxidation and adsorption processes to precipitate dissolved Fe^{2+} and Mn^{2+} ions into insoluble Fe and Mn oxides which are removed at the filter bed.
- This low-cost and eco-friendly technology shows high potentials for application to removal of toxic inorganic As species from contaminated water. However, the intricate mechanisms of biological Fe and Mn oxidation as well as the integration of this system with As removal are not fully understood.

References

1. Mouchet, P., *J. Am. Water Works Assoc.* **84**(4), 158–167 (1992).
2. Tamura, T., Tsunai, T., Ishimaru, Y., and Nakata, A., *Suido Kyokai Zasshi* (*J. Japan Water Works Assoc.*) **68**, 1–13 (1999) (in Japanese).
3. Kojima, S., *Yosui to Haisui (J. Water Waste)* **14**, 709–715 (1972) (in Japanese).
4. Tanimoto, K., *Suido Kyokai Zasshi* (*J. Japan Water Works Assoc.*) **213**, 19–21 (1952) (in Japanese).
5. Czekalla, C., Mevius, W., and Hanert, H. H., *Water Supply* **3**, 111–123 (1985).

6. Bourgine, F. P., Gennery, M., Chapman, J. I., Kerai, H., Green, J. G., Rap, R. J., Ellis, S., and Gaumard, C., *Water Environ. J.* **8**, 379–391 (1994).
7. Burger, M. S., Krentz, C. A., Mercer, S. S., and Gagnon, G. A., *J. Water Supply Res. Technol. Aqua* **57**, 351–359 (2008).
8. Fujikawa, Y., Ikejima, M., Yukimoto, M., Tamura, T., Takada, K., and Hamasaki, T., *Yosui to Haisui (J. Water Waste)* **50**, 277–287 (2008) (in Japanese).
9. Li, D., Zhang, J., Wang, H., Yang, H., and Wang, B., *J. Water Supply Res. Technol. Aqua* **54**, 15–24 (2005).
10. Olańczuk-Neyman, K. and Bray, R., *Pol. J. Environ. Stud.* **9**, 91–96 (2000).
11. Pacini, V. A., Ingallinella, A. M., and Sanguinetti, G., *Water Res.* **39**, 4463–4475 (2005).
12. Vandenabeele, J., de Beer, D., Germonpré, R., Van de Sande, R., and Verstraete, W., *Water Res.* **29**, 579–587 (1995).
13. Vandenabeele, J., de Beer, D., Germonpré, R., and Verstraete, W., *Microb. Ecol.* **24**, 91–108 (1992).
14. van Veen, W. L., Mulder, E. G., and Deinema, M. H., *Microbiol. Rev.* **42**, 329–356 (1978).
15. Peitchev, T. and Semov, V., *Water Sci. Technol.* **20**, 173–178 (1988).
16. Sahabi, D. M., Takeda, M., Suzuki, I., and Koizumi, J., *J. Biosci. Bioeng.* **107**, 151–157 (2009).
17. Štembal, T., Markić, M., Ribičić, N., Briški, F., and Sipos, L., *Process Biochem.* **40**, 327–335 (2005).
18. Suzuki, I., Takeda, M., and Koizumi, J., *Seibutsu Kogaku Kaisi* **86**, 605–607 (2008) (in Japanese).
19. Frischherz, H., Zibuschka, F., Jung, H., and Zerobin, W., *Water Supply* **3**, 125–136 (1985).
20. Fujikawa, Y., Hamasaki, T., Sugahara, M., Minami, A., Tani, S., and Tonokai, K., *Yosui to Haisui (J. Water Waste)* **50**, 105–113 (2008) (in Japanese).
21. Hanert, H. H., *The Prokaryotes*, 3rd Ed., **7**, eds. Dworkin, M., Falkow, S., Rosenberg, E., Schleifer, K.-H., and Stackebrandt, E., Chapter 14.7 "The genus *Gallionella*," (Springer, New York) pp. 990–995 (2006).
22. Spring, S., *The Prokaryotes*, 3rd Ed., **5**, eds. Dworkin, M., Falkow, S., Rosenberg, E., Schleifer, K.-H., and Stackebrandt, E., Chapter 3.2.11 "The genera *Leptothrix* and *Sphaerotilus*," (Springer, New York) pp. 758–777 (2006).

23. Moore, R. L., *Annu. Rev. Microbiol.* **35**, 567–594 (1981).
24. Poindexter, J. S., *The Prokaryotes*, 3rd Ed., **5**, eds. Dworkin, M., Falkow, S., Rosenberg, E., Schleifer, K.-H., and Stackebrandt, E., Chapter 3.1.3 "Dimorphic prosthecate bacteria: The genera *Caulobacter*, *Asticcacaulis*, *Hyphomicrobium*, *Pedomicrobium*, *Hyphomonas* and *Thiodendron*," (Springer, New York) pp. 72–90 (2006).
25. Tyler, P. A., *Antonie van Leeuwenhoek* **36**, 567–578 (1970).
26. Hirsch, P., *Bergey's Manual of Systematic Bacteriology*, **3**, eds. Staley, J. T., Bryant, M. P., Pfennig, N., and Holt, J. G. "Sheathed bacteria—genus *Clonothrix*," (Williams and Wilkins, Baltimore) pp. 2008–2009 (1989).
27. Hirsch, P., *Bergey's Manual of Systematic Bacteriology*, **3**, eds. Staley, J. T., Bryant, M. P., Pfennig, N., and Holt, J. G. "Sheathed bacteria—genus *Crenothrix*," (Williams and Wilkins, Baltimore) pp. 2006–2008 (1989).
28. Hirsch, P., *The Prokaryotes*, 3rd Ed., **7**, eds. Dworkin, M., Falkow, S., Rosenberg, E., Schleifer, K.-H., and Stackebrandt, E., Chapter 14.6 "The genus *Toxothrix*," (Springer, New York) pp. 986–989 (2006).
29. Krul, J. M., Hirsch, P., and Staley, J. T., *Antonie van Leeuwenhoek* **36**, 409–420 (1970).
30. Hanert, H. H., *The Prokaryotes*, 3rd Ed., **7**, eds. Dworkin, M., Falkow, S., Rosenberg, E., Schleifer, K.-H., and Stackebrandt, E., Chapter 14.11 "The genus *Siderocapsa* (and other iron- and manganese-oxidizing eubacteria)," (Springer, New York) pp. 1005–1015 (2006).
31. Emerson, D., Garen, R. E., and Ghiorse, W. C., *Arch. Microbiol.* **151**, 223–231 (1989).
32. Maki, J. S., Tebo, B. M., Palmer, F. E., Nealson, K. H., and Staley, J. T., *FEMS Microbiol. Lett.* **45**, 21–29 (1987).
33. Gouzinis, A., Kosmidis, N., Vayenas, D., and Lyberatos, G., *Water Res.* **32**, 2442–2450 (1998).
34. Suzuki, I., Takezaki, J., Kusumoto, H., Makita, H., Takeda, M., and Koizumi, J., *Proc. 11th Int'l. Symp. Microb. Ecol.* ISME, A77 (2006).
35. Takeda, M., Kamagata, Y., Ghiorse, W. C., Hanada, S., and Koizumi, J., *Int. J. Syst. Evol. Microbiol.* **52**, 895–900 (2002).
36. Straub, K. L., Benz, M., Schink, B., and Widdel, F., *Appl. Environ. Microbiol.* **62**, 1458–1460 (1996).
37. Siering, P. L. and Ghiorse, W. C., *Appl. Environ. Microbiol.* **63**, 644–651 (1997).

38. Japanese Ministry of Environment, 2006 *Nationwide Water Quality Survey of Groundwater* (2007) (in Japanese).

39. Fujikawa, Y., Yoneda, D., Minami, A., Yashima, H., Tonokai, K., Tani, S., Fukui, M., Hamasaki, T., and Sugahara, M., *Chinese J. Geochem.* **25**, 113 (2006).

40. Sahabi, D. M., Takeda, M., Suzuki, I., and Koizumi, J., *J. Hazard. Mater.* **168**, 1310–1318 (2009).

41. Zouboulis, A. I. and Katsoyiannis, I. A., *Environ. Int.* **31**, 213–219 (2005).

42. Katsoyiannis, I. A. and Zouboulis, A. I., *Water Qual. Res. J. Canada* **41**, 117–129 (2006).

43. Katsoyiannis, I. A. and Zouboulis, A. I., *Water Res.* **38**, 17–26 (2004).

44. Pokhrel, D. and Viraraghavan, T., *J. Environ. Manage.* **90**, 1956–1961 (2009).

Chapter 3

REMOVAL OF SELENIUM FROM WASTEWATER USING A SELENIUM-REDUCING BACTERIUM

Michihiko Ike and Satoshi Soda
Division of Sustainable Energy and Environmental Engineering, Graduate School of Engineering, Osaka University, 2-1 Yamada-oka, Suita, Osaka 565-0871, Japan
ike@see.eng.osaka-u.ac.jp

3.1 Introduction

Selenium (Se) is widely used in various industries such as glass, pigment, pesticide, stainless steel, and photoelectric cell production.[1] Wastewater from these industries contains considerable amounts of soluble Se, selenate (SeO_4^{2-}) and selenite (SeO_3^{2-}). Furthermore, irrigation of Se-rich fields causes an increase in the concentration of soluble Se in drainage water. High concentrations of soluble Se are acutely and chronically toxic to organisms inhabiting aquatic environments.[2]

In Japan, severe Se contamination has not been reported to date. Nevertheless, the high probability of contamination remains

Handbook of Metal Biotechnology: Applications for Environmental Conservation and Sustainability
Edited by Michihiko Ike, Mitsuo Yamashita, and Satoshi Soda

www.panstanford.com

because Japan refines and consumes the largest amount of Se in the world. Wastewater from some industries in Japan actually contains high concentrations of soluble Se, e.g., 620 mg Se/L from the Se-compounds industry[18] and 20–60 mg Se/L from the copper refining industry.[19,20] Japan's environmental quality standards for water pollution were amended in 1993 for Se; maximum concentrations of Se permissible in industrial wastewater were set at 0.1 mg Se/L. However, because appropriate practical methods for removing soluble Se from industrial wastewater have not yet been established, provisional standards have been set for some industrial wastewater (e.g. 0.3 mg Se/L for Se industry) until January 2009.

At present, physicochemical methods such as chemical precipitation, catalytic reduction, and ion exchange are mainly used for removing Se from wastewater. These methods are effective for removing selenite, but not selenate.[3] Therefore, their application is limited. The electrochemical reduction process of selenate to selenite is applicable for pretreatment. These methods commonly present the problem of high costs. Consequently, it is desirable to develop cost-effective Se removal processes that are effective not only for selenite but also for selenate. An attractive alternative for cost-effective Se removal might be biological treatment using Se-reducing microbes. It is known that microbial reduction of selenate into elemental selenium (Se^0) via selenite plays an important role in detoxification of soluble Se in the natural environment. Because elemental Se is of little or no toxicity and is easily removed from the aqueous phase because of its insoluble characteristics,[4,5,13] this reductive process might be applied to develop wastewater treatment systems for detoxification and removal of soluble Se, especially selenate. In fact, some laboratory-scale and pilot-scale biological Se treatment systems have been reported for wastewater containing low concentrations of soluble Se, such as agricultural drainage water[6,7,22] and refinery wastewater.[8,11] No attempt has been made to develop a biological treatment process that is applicable to the high concentrations of soluble Se in industrial wastewater.

Our research group has isolated a selenate-reducing bacterium, *Bacillus selenatarsenatis* SF-1[10,17] and has investigated factors affecting its soluble Se removal capability.[13,15] Based on these studies, strain SF-1 is considered a promising candidate for use in the treatment of wastewater containing high concentrations of soluble Se because this bacterial strain can tolerate high concentrations of selenate (up to about 150 mg Se/L); it can reduce selenate into selenite efficiently, and subsequently, into elemental Se. In this study, laboratory-scale bioreactors using *B. selenatarsenatis* SF-1 were reviewed to investigate its applicability for removal of high concentrations of soluble Se.

3.2 Selenate- and Selenite-Reducing Bacterium *B. Selenatarsenatis* SF-1

A gram-positive, selenate-reducing bacterium, strain SF-1, was isolated from Se-contaminated sediment (4.8 mg Se/kg) collected from an effluent drain that had been receiving Se-containing discharge from a glass manufacturing plant.[10] Plating of the culture on a nutrient agar plate containing selenate caused the formation of uniform red colonies, suggesting the accumulation of amorphous Se. Based on the physiological, phylogenetic, and molecular evidence presented herein, the strain represents a novel species of the genus *Bacillus: B. selenatarsenatis.*[17]

A large amount of inoculum of this facultative anaerobe can be prepared easily under aerobic conditions, which is advantageous for wastewater treatment purposes. The bacterial strain can grow with lactate as an electron donor and a carbon source while using selenate as an electron acceptor under anoxic conditions. In this anaerobic respiration (selenate respiration), selenate is transformed reductively into selenite, then into elemental Se. Se respiration on lactate by bacteria of some kinds is expressed theoretically as follows.[5,10,12]

$$\text{Lactate}^- + 2\text{SeO}_4^{2-} \rightarrow \text{acetate}^- + 2\text{SeO}_3^{2-} + \text{HCO}_3^- + \text{H}^+$$
$$(\Delta Gf^0 : -343.1 \text{ kJ/mol of lactate}) \qquad (3.1)$$

$$\text{Lactate}^- + \text{SeO}_3^{2-} + \text{H}^+ \rightarrow \text{acetate}^- + \text{Se}^0 + \text{HCO}_3^- + \text{H}_2\text{O}$$
$$(\Delta Gf^0 : -467.4 \text{ kJ/mol of lactate}) \qquad (3.2)$$

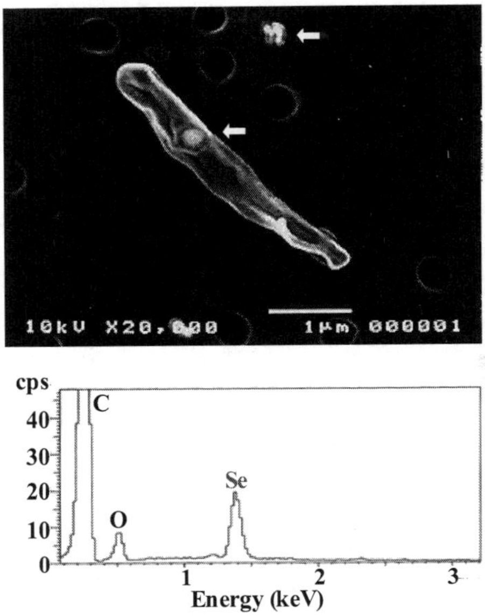

Figure 3.1. SEM observation of a selenium-accumulating cell (a) and EDS analysis of a particle found in a cell of *B. selenatarsenatis* SF-l (b). Arrows indicate elemental selenium.

As shown in Fig. 3.1, scanning electron microscope (SEM) observation of the SF-1 culture resulting from the selenate reduction test revealed electron-dense particles with diameters of approximately 200–300 nm both inside and outside the cells. These particles were identified as Se using energy dispersive X-ray spectrometry (EDS) analysis. Because Se^0 is insoluble, it can be removed easily from an aqueous phase by filtration or centrifugation.[13] However, whether the Se particles first occurred inside the cells and were subsequently expelled or they were deposited both inside and outside the cells could not be clarified. To date, at least three genes related to selenate reduction have been found in *B. selenatarsenatis* SF-1 — *srdA*, *srdB*, and *srdC* — which respectively encode a catalytic subunit with a molybdopterin binding site, a subunit containing four [4Fe–4S] clusters, and a membrane anchor protein subunit.[21]

For realization of such systems, it is important to elucidate the mechanisms of this Se reduction process. Selenate- and selenite-reducing microorganisms should be identified and fully characterized. Optimal pH and temperature for selenate reduction are, respectively, about 8.0 and 30°C. Pyruvate, lactate, sucrose, and fructose are particularly effective substrates for promoting selenate reduction. Furthermore, synthetic sewage mainly containing meat extract could support the effective growth of *B. selenatarsenatis* SF-1 and its selenate reduction activity. However, none of the tested carbon sources was effective in selenite reduction except for synthetic sewage. The selenate-reducing rate by *B. selenatarsenatis* SF-1 without other rate-limiting factors increases linearly with selenate concentration up to about 400 mg Se/L.[13] The specific selenite-reduction rate also increases with selenite concentrations up to about 160 mg Se/L.[13] It is important to note that the first-order kinetic constant of this bacterium for selenate reduction (2.9 $\times$ 10^{-11} L/cells/h) is much larger than that for selenite reduction (5.5 $\times$ 10^{-13} L/cells/h). On the other hand, Rege *et al.*[11] reported that the estimated parameter value for expressing the selenate-reducing rate (0.93 L/gbiomass/mol/min) was considerably lower than that for the selenite-reducing rate (4.11 L/gbiomass/mol/min) of a denitrifying bacterial consortium in a sequencing batch reactor. Reportedly, selenite reduction proceeded more rapidly than the selenate reduction and the rate-limiting step in the detoxification of soluble Se lies in the conversion of selenate into selenite in Se-polluted[4] and non-Se-polluted aquatic environments.[14] The yield of the bacterial cells by selenate reduction was estimated as 2.2 $\times$ 10^9 cells/mg-Se.[16] In fact, *B. selenatarsenatis* SF-1 showed slight growth by selenite reduction, but it was unstable and not reproducible.[10]

3.3 Lab-Scale Sequencing Batch Reactor

The possibility of constructing a soluble Se removal process using *B. selenatarsenatis* SF-1 was investigated using an anaerobic culture operated in a sequencing batch mode at 30°C.[15] Typical results are presented in Fig. 3.2. Selenate and selenite were measured using ion chromatography.

Strain SF-1 transformed about 40 mg Se/L of selenate into elemental Se within a 24-h cycle over at least 10 cycles. Although 8–25 mg Se/L of selenite was transiently accumulated in each

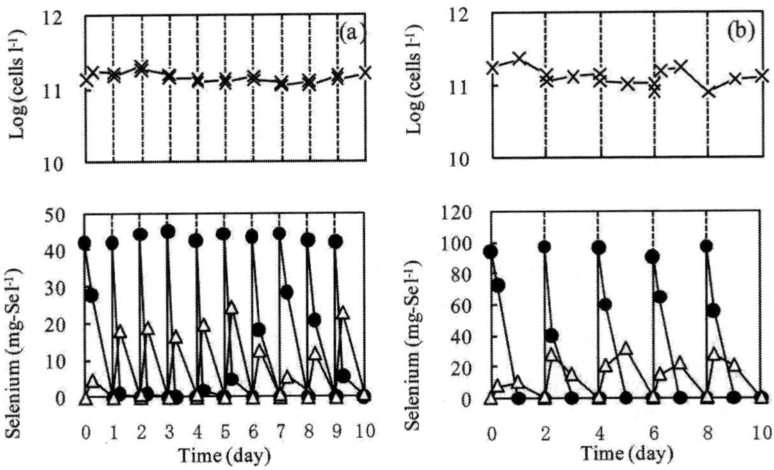

Figure 3.2. Sequencing batch test using SF-1 for selenate removal. The selenate concentration was about 40 mg Se/L (a) or about 80 mg Se/L (b). During each batch cycle, a total of 2 mL (a) and 5 mL (b) of the culture was withdrawn and used for routine analysis. Selenate (closed circle), selenite (open triangle), and cells (cross) are shown.

cycle, it was subsequently reduced to insoluble Se during the cycle period. Selenate removal and selenite accumulation were observed and the effluent turned deep red in all runs, indicating formation of amorphous elemental Se. The cell density remained stable at $1.2 \times 10^{11} - 2.1 \times 10^{11}$ cells/L. Furthermore, *B. selenatarsenatis* SF-1 was able to transform about 80 mg Se/L of selenate stably into elemental Se within a 48-h cycle over at least five cycles (Fig. 3.2b). During these experiments, a specified amount of the culture broth was withdrawn each day for analysis and cell counting. The hydraulic and cell retention time of the sequencing batch experiment with 24-h cycle was 10 d; that with a 48-h cycle was 8 d. In both cases, the cell density remained stable throughout the experimental period. For that reason, the amount of cells withdrawn was estimated to balance the cell growth. These results indicate that it will be possible to use strain SF-l cells repeatedly for removing soluble Se at a load of approximately 40 g Se (as selenate)/m^3/d in an anaerobic sequencing batch mode.

3.4 Lab-Scale Continuous Reactor

A continuous bioreactor using *B. selenatarsenatis* SF-1 was constructed and operated as a lab-scale chemostat (0.5 L) to investigate its applicability to treatment of wastewater containing highly concentrated Se (selenate).[16] Each experimental run was conducted at least 2.5 times longer than the corresponding cell retention time at 2.9–95.2 h at 30°C. In all runs, Se concentrations and cell density in the bioreactor stabilized after 20–30 h, indicating that steady states were obtained. Because it is known that *B. selenatarsenatis* SF-1 cannot transform Se into gaseous forms,[10] selenate reduction products were considered to be selenite and elemental Se in this study. Bacterial cell density was maintained at 10^{10}–10^{12} cells/L.

Steady-state data are expressed as averages of values obtained from samples at the steady state (at least three times for each run). They are depicted in Fig. 3.3 as a function of the cell retention

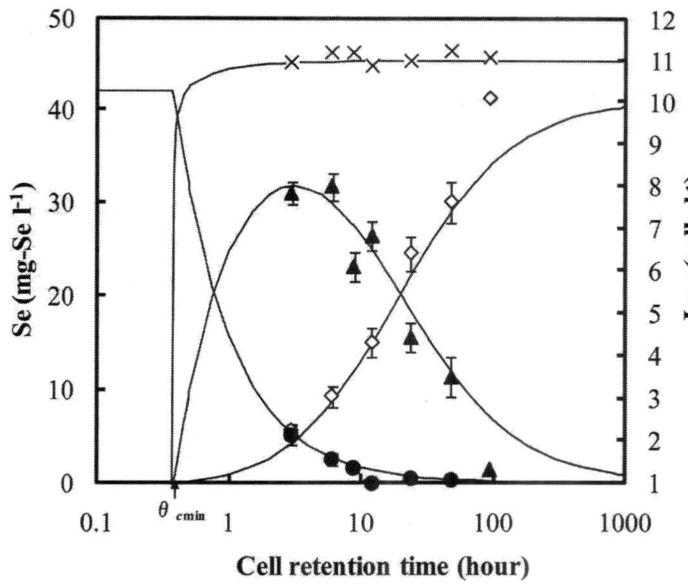

Figure 3.3. Effect of cell retention time on Se and bacterial concentrations in the bioreactor using *Bacillus* sp. SF-1 at steady states. Experimental data for selenate (closed circle), selenite (closed triangle), elemental Se (open diamond), and total bacterial cell (cross) concentrations are shown. Minimum cell retention time θ_{min} is about 0.37 h.

time. As the figure shows, the selenate concentration in the effluent decreased along with increased cell retention time. Concentration fell below detection limits (0.05 mg Se/L) at retention times of more than 12 h. More than 85% of the selenate was removed even with the extremely short cell retention time of 2.9 h (14 mg Se/L/h of volumetric Se-loading rate), although removal of soluble Se (selenate + selenite) was less than 20%. At shorter cell retention times, the removed selenate was accumulated mainly as selenite. However, its concentration decreased when the cell retention time lengthened. At the cell retention time of 95.2 h (0.45 mg Se/L/h of volumetric Se-loading rate), almost all soluble Se was removed (>99%) and transformed into insoluble elemental Se: highly effective soluble Se removal was achieved.

Although the selenate removal efficiency of the lab-scale bioreactor was maintained at high levels even with shorter retention times, selenite tended to accumulate at higher concentrations. Because various physicochemical methods including chemical precipitation, catalytic reduction, and ion exchange are effective for removing selenite,[3] the combined use of the bioreactor and such physicochemical methods can serve as a strategy for high-rate treatment of selenate-containing wastewater. Another strategy for high-rate treatment might be the combined use of the selenate-reducing bacterium SF-1 and selenite-reducing bacteria. Numerous microorganisms are capable of reducing selenite to elemental Se. In fact, selenite is more readily reduced than selenate in the natural environment.[4,11,14] Moreover, previous reports show that normal, biological anaerobic wastewater treatment processes are available to remove selenite.[8,9] Bioaugmentation — inoculation of specific microbes into the environment — of *B. selenatarsenatis* SF-1 into an existing anaerobic wastewater treatment process is one method for such combined treatment. Another way is the use of the bioreactor using *B. selenatarsenatis* SF-1, with subsequent use of existing anaerobic wastewater treatment.

3.5 Conclusions and Outlook

The biological treatment process using *B. selenatarsenatis* SF-1 is useful for detoxification and removal of soluble Se from

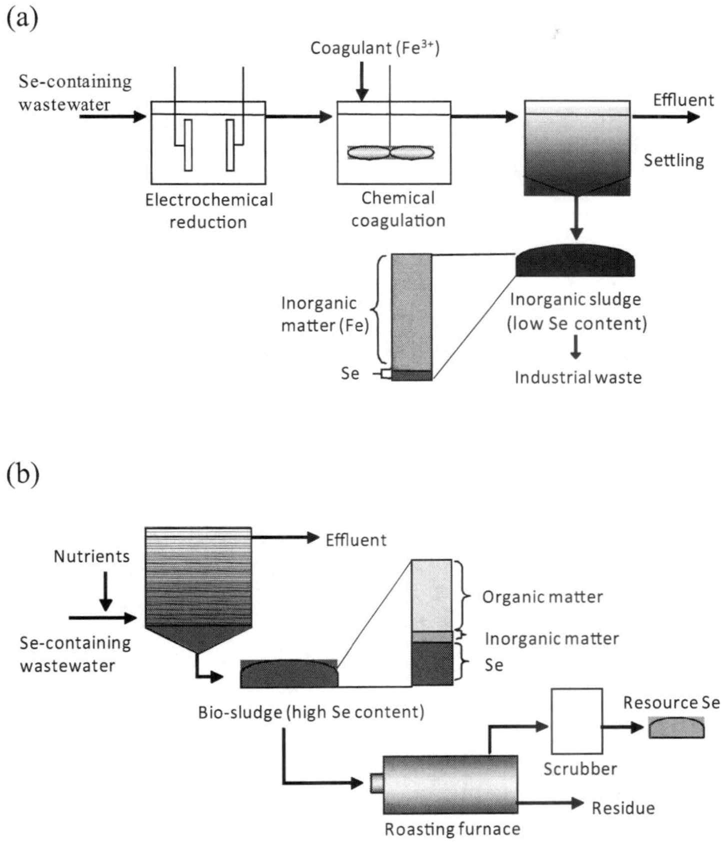

Figure 3.4. Schematic treatment systems for selenium-containing wastewater. (a) A typical physico-chemical system using an electrochemical reduction process and a coagulation process, (b) a new system using a biological reduction process and a roasting furnace for selenium recovery. See also Color Insert.

Se-containing wastewater in sequencing batch and continuous modes. Furthermore, this process is advantageous for conventional physicochemical processes such as coagulation and adsorption because it can recover Se as a resource.

A typical physicochemical process and a proposed biological process for Se removal from wastewater are portrayed in Fig. 3.4. The effluent standard for Se can be achieved through coagulation with $FeCl_3$ following electrochemical reduction of selenate to selenite, but the resultant inorganic sludge with >99% Fe should be disposed of as industrial waste (Fig. 3.4a). On the other hand, the anaerobic bioreactor proposed in this study merely requires electricity for stirring and electron donors (carbon sources) and nutrients for selenate-reducing and selenite-reducing bacterial strains (Fig. 3.4b). The resulting excess biomass contains insoluble elemental Se at >5% dry-weight. Through incineration of the dry biomass, organic carbon is removed to air as CO_2 and vaporized Se is recoverable as a resource through the alkali scrubber process. Biological processing with selenate-reducing and selenite-reducing bacteria for Se-containing wastewater presents the attractive possibility of use as a Se recovery process without sludge disposal.

Until now, our experiments have been performed using a pure culture under sterilized conditions. To realize a practical Se-removal system, further experiments must be conducted under non-sterilized conditions.

Acknowledgments

A part of this work was conducted as the Regional New Consortium Project commissioned by the Ministry of Economy, Trade and Industry, Japan (METI), and the Kansai Bureau Economy, Trade and Industry (METI-KANSAI). We would like to thank H. Takahashi and N. Iwasaki (Shinko Chemical Corp.), and K. Uesugi and M. Eguchi (Organo Co.) for their helpful suggestions.

References

1. Haygarth, P. M., in: *Selenium in the Environment*, eds. Frankenberger Jr., W.T. and Benson, S. (Marcel Dekker, New York) pp. 1–7 (1994).
2. Gillespie, R. B. and Baumann, P. C., *Trans. Am. Fish. Soc.* **115**, 208 (1986).
3. Lee, E. W., in: *Biotreatment of Agricultural Wastewater*, eds. Huntley, M. E. (CRC Press, Boca Raton, FL) pp. 33–45 (1989).
4. Maiers, D. T., Wichlacz, P. L., Thompson, D. L., and Bruhn, D. F., *Appl. Environ. Microbiol.* **54**, 2591 (1988).
5. Oremland, R. S., Blum, J. S., Culbertson, C. W., Vesscher, P. T., Miller, L. G., Dowdle, P., and Strohmaier, F. E., *Appl. Environ. Microbiol.* **60**, 6311 (1994).

6. Cantafio, A. W., Hagen, K. D., Lewis, G. E., Bledsoe, T. L., Nuran, K. M., and Macy, J. M., *Appl. Environ. Microbiol.* **62**, 3298 (1996).
7. Gerhardt, M. B., Green, F. B., Newman, R. D., Lundquist, T. J., Tresan, R. B., and Oswald, W. J., *Res. J. WPCF* **63**, 799 (1991).
8. Lawson, S. and Marcy, J. M., *Appl. Microbiol. Biotechnol.* **43**, 762 (1995).
9. Garbisu, C., Alkorta, I., Carlson, D. E., Leighton, T., and Buchanan, B. B., *Microbiol. SEM* **13**, 437 (1997).
10. Fujita, M., Ike, M., Nishimoto, S., Takahashi, K., and Kashiwa, M., *J. Ferment. Bioeng.* **83**, 517 (1997).
11. Rege, M. A., Yonge, D. R., Mendoza, D. P., Petersen, J. N., Bereded-Samuel, Y., Johnstone, D. L., Apel, W. A., and Barnes, J. M., *Biotechnol. Bioeng.* **62**, 479 (1999).
12. Blum, J. S, Bindi, A. B., Buzzelli, J., Stolz, J. F., and Oremland, R. S. *Arch. Microbiol.* **171**, 19–30 (1998).
13. Kashiwa, M., Nishimoto, S., Takahashi, K., Ike, M., and Fujita, M., *J. Biosci. Bioeng.* **89**, 528 (2000).
14. Ike, M., Takahashi, K., Fujita, T., Kashiwa, M., and Fujita, M., *Water Res.* **34**, 3019 (2000).
15. Kashiwa, M., Ike, M., Mihara, H., Esaki, N., and Fujita, M., *Biofactors* **14**, 261 (2001).
16. M. Fujita, M. Ike, M. Kashiwa, R. Hashimoto, and S. Soda, *Biotechnol. Bioeng.* **80**, 755 (2002).
17. Yamamura, S., Yamashita, M., Fujimoto, N., Kuroda, M., Kashiwa, M., Sei, K., Fujita, M., and Ike, M., *Int. J. Sys. Evol. Microbiol.* **57**, 1060 (2007).
18. Shimomura, K., Hatsugai, R., Iizuka, S., and Eguchi, N., Yokohama-shi-kankyo-kagaku-kenkyu-jo-hou, **21**, 119 (1997) (in Japanese).
19. Yarita, K. and Yukawa, Y., *J. Water Waste* **42**,733 (2000) (in Japanese).
20. Kawabata, T., Imai, T, and Asai, K., *Bull. Jpn Mining Ind. Assoc.* **50**, 6 (1997) (in Japanese).
21. Kuroda, M., Yamashita, M., Miwa, E., Fujimoto, N., Ono, H., Nagano, K., Sei, K., and Ike, M., *J. Bacteriaol.* **193**, 2142 (2011).

Chapter 4

MICROBIAL REDUCTION AND MOBILIZATION OF ARSENIC FOR BIOREMEDIATION OF CONTAMINATED SOIL

Shigeki Yamamura

Center for Regional Environmental Research, National Institute for Environmental Studies, 16-2 Onogawa, Tsukuba, Ibaraki, 305-8506, Japan

yshige@nies.go.jp

4.1 Introduction

Arsenic (As), commonly recognized as a toxic metalloid, is widely distributed in nature,[1] and its background concentration in soils is generally less than 15 mg As kg^{-1}.[2] However, the distribution of As can vary depending on the release of naturally occurring As through activities such as volcanism, hydrothermal activity, and weathering of parent materials containing As-bearing minerals. Anthropogenic activities, such as air emissions, soil amendments, mining operations, and wood preservation, have also led to elevated

Handbook of Metal Biotechnology: Applications for Environmental Conservation and Sustainability
Edited by Michihiko Ike, Mitsuo Yamashita, and Satoshi Soda

www.panstanford.com

As levels.[2,3] Chronic exposure to As has caused a wide variety of adverse health effects including dermatological conditions and various cancers.[4] Recent studies also indicate that As is a potent endocrine disruptor, potentially relevant to human developmental problems and disease risk.[5,6] Health problems associated with exposure to As are a global issue, most notably in Southeast Asia. In Bangladesh alone, it is estimated that more than 57 million people rely on drinking water that contains As in excess of the World Health Organization (WHO) standard.[7]

As has also become a prevalent soil contaminant throughout the world.[3,8] Accumulated As in soils can extensively pollute ground and surface water as a result of leaching into downstream areas, increasing animal and human health risks. The Japanese Ministry of the Environment enforced the Soil Contamination Countermeasures Law in 2003 to address issues caused by harmful substances including As. Thereby, designated areas where soil standards are not met require appropriate management, and the remediation of contaminated sites has become an important environmental concern. The techniques that are currently available for the treatment of As-contaminated soils are mainly soil replacement, containment, and solidification/stabilization,[9,10] which are expensive and/or require long-term monitoring and management to prevent the risk of leaching in the future. In addition, soil replacement can produce large amounts of secondary pollutants via disposal of contaminated soils, which have further costs and require a great deal of attention for subsequent treatment.

Microbial-based technologies for remediation, which are cost effective in comparison to physicochemical treatments, have primarily focused on degradation of organic contaminants. There have, however, been some investigations into the use of microbial processes for the remediation of soils contaminated with metals and metalloids, including As. Microorganisms can alter speciation of metal(loid)s, which plays an important role in mobilization and immobilization. This chapter seeks to highlight mechanisms of microbial As mobilization and its potential for use in bioremediation of As-contaminated soils.

4.2 Microbial As Mobilization

Although As can exist in both inorganic and organic forms, inorganic arsenate (As(V)) and arsenite (As(III)) predominate in soils.[11,12] As(V) is present as negatively charged oxyanions ($H_2AsO_4^-/HAsO_4^{2-}$) at nearly neutral pH and strongly binds to soil minerals including Fe and Al (hydr)oxides.[2,12] By contrast, most As(III) exists as uncharged $H_3AsO_3^0$. Therefore, As(III) is less adsorptive and more mobile than As(V) in most environments,[2,12] although it can be specifically adsorbed onto Fe oxides.[13] Under aerobic conditions, As(V) is found to be the major species and is immobilized in soil minerals, whereas anaerobic conditions lead to mobilization mainly as As(III) into an aqueous phase.[14]

Anaerobic microbes play a key role in the mobilization of As by either a direct or an indirect mechanism (Fig. 4.1). The former is the reduction of As(V) associated with the solid phase to mobile As(III). Microorganisms can catalyze As redox transformations and affect As mobility,[15,16] that is, microbial As(V) reduction directly mediates dissolution of As from As(V)-laden minerals.[17–19] The latter is the reductive dissolution of Fe(III) minerals containing As. Fe(III)-reducing bacteria reduce insoluble Fe(III) (hydr)oxides to soluble Fe(II), resulting in indirect dissolution of associated As.[20]

Microbial mobilization of As from soils and sediments can be a significant route of human exposure via contamination of natural waters. However, under well-controlled conditions, such microbial processes can be an attractive, cost-effective remediation alternative for promoting As removal from contaminated soils.

4.3 Role of Dissimilatory As(V)-Reducing Bacteria in As Mobilization

When microbial As mobilization is utilized as a strategy for As removal from contaminated soil, the direct mechanism, that is, As(V) reduction, might be an important pathway because the indirect mechanism via Fe(III) reduction cannot mobilize As associated with soil minerals other than Fe(III), such as Al (hydr)oxides

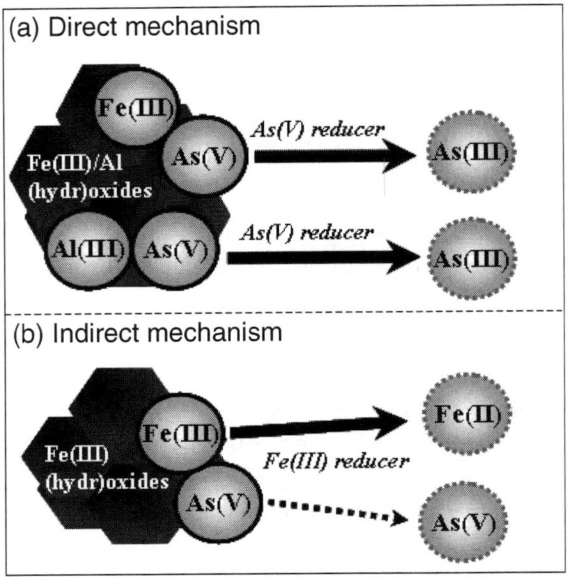

Figure 4.1. Possible mechanisms of As mobilization by anaerobic microbes. (a) Direct mechanism; release of As(III) from either Fe(III) or Al minerals via reduction of As(V) to As(III) mediated by As(V) reducers. (b) Indirect mechanism; release of As(V) from Fe(III) minerals via reduction of Fe(III) to Fe(II) mediated by Fe(III) reducers. Note that some of the released As(III) and As(V) can re-adsorb to unreacted Fe(III) and/or Fe(II)-bearing minerals.

(Fig. 4.1). In addition, under experimental conditions, the rate and extent of As mobilization from amorphous Fe(III) precipitates was considerably greater in the presence of both microbial As(V) and Fe(III) reduction than with only Fe(III) reduction (Fig. 4.2).[21] Recent studies have also reported that complex pathways of Fe biotransformation control As mobilization and immobilization in Fe (hydr)oxides, and Fe(II)-bearing minerals produced via microbial Fe(III) reduction sometimes promote As immobilization rather than mobilization.[22,23] Thus, several studies have indicated that reduction of As(V) might be the key process of As mobilization,[21,24] although the Fe mineral content is strongly linked to the retention of As in soils,[25] and reductive dissolution of Fe(III) minerals can

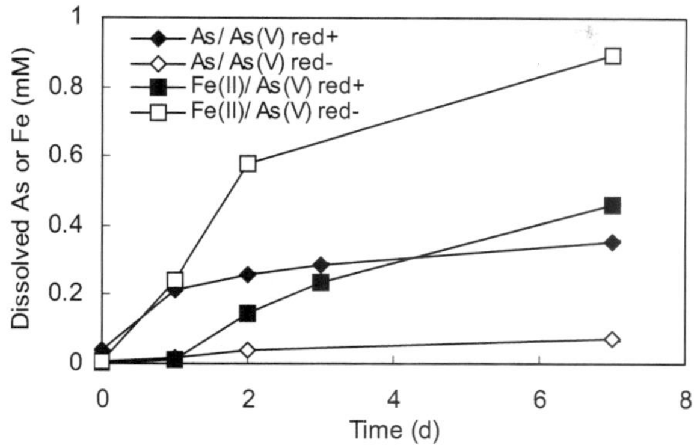

Figure 4.2. Reductive dissolution of As and Fe from As(V)-laden Fe(III) precipitates as mediated by As(V)- and Fe(III)-reducing bacteria. As(V) red+; cultures containing solid-phase As(V) and Fe(III) were incubated with *Bacillus selenatarsenatis*, which is able to reduce both As(V) and Fe(III). As(V) red−; identical cultures were incubated with *Bacillus jeotgali*, which is able to reduce Fe(III), but not As(V).

support the direct pathway by increasing microbial accessibility to solid-phase As(V).[21]

Microbial reduction of As(V) to As(III) can occur through two principal processes (Fig. 4.3).[15,16,26] One is dissimilatory reduction (respiration), where dissimilatory As(V)-reducing bacteria (DARB) utilize As(V) as a terminal electron acceptor for anaerobic respiration. DARB can achieve growth via the oxidation of various organic or inorganic electron donors, resulting in As(V) reduction. In addition, they can reduce either aqueous or solid-phase As(V) because respiratory As(V) reductase is located on the periphery of the cells.[16–19] The other process is involved in the detoxification mechanism. A variety of microbes, both aerobic and anaerobic, have an As resistance system comprising cytoplasmic As(V) reductase and As(III)-specific transporters to expel As from the cells. In contrast to DARB, such As-resistant bacteria cannot gain energy from the process and can only reduce aqueous As(V) that has entered the cells.[16,27] Thus, DARB play an important role in As mobilization and might be desirable agents for As bioremediation, because As(V) is often found to be the predominant species in As-contaminated soils.[28–30]

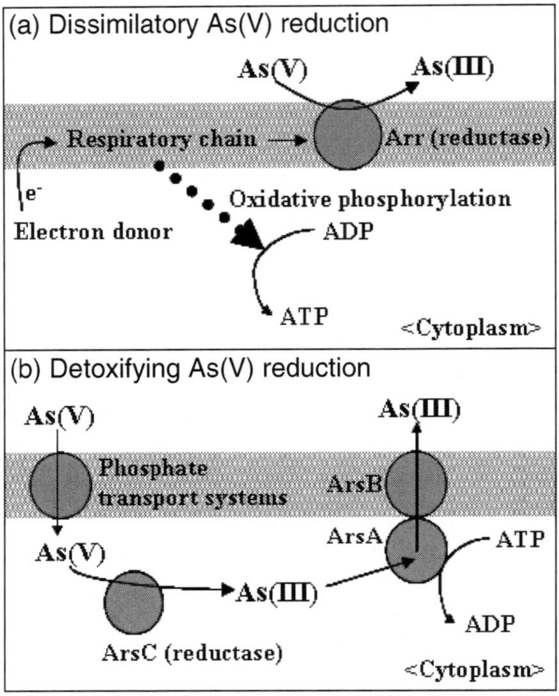

Figure 4.3. Pathways of bacterial As(V) reduction. (a) Dissimilatory reduction; extracellular As(V) is reduced by the terminal reductase (Arr) of the respiratory chain, coupled with oxidative phosphorylation. (b) Detoxifying reduction; intracellular As(V), accumulated by phosphate transport systems, is reduced by the cytoplasmic reductase (ArsC). As(III) is pumped out of the cells by the membrane As(III) efflux proteins (ArsB) coupled with an ATPase (ArsA).

4.4 Effect of the Electron Shuttle on Microbial As Mobilization

Humic substances, which represent the majority of organic matter in aquatic and soil environments, can be reduced by many microbes as well as various metal(loid)s including Fe(III) and As(V), because they contain quinone moieties as the dominant electron acceptor.[31,32] Reduced humic substances can transfer electrons to various Fe(III) minerals and are re-oxidized concomitantly with the

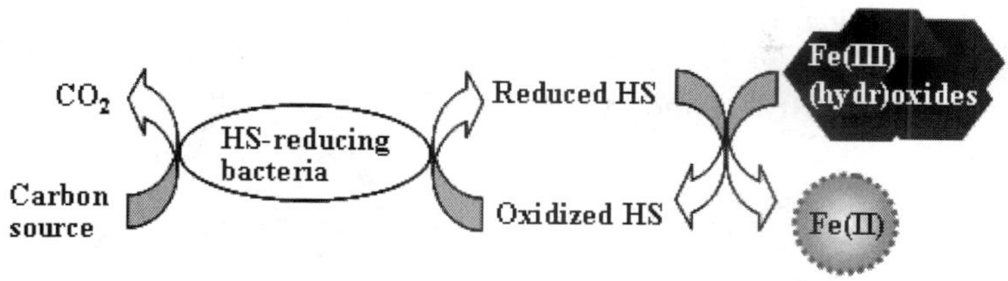

Figure 4.4. Role of humic substances (HS) as electron-shuttling compounds. Oxidized HS can be utilized as electron acceptors for microbial oxidation of organic carbon sources. Reduced HS can act as chemical reductants for Fe(III) reduction with concomitant re-oxidation to oxidized HS.

chemical reaction.[33] This microbial reduction of quinone moieties and extracellular electron-shuttling allows for an indirect reduction of insoluble Fe(III) and enhances the rates of microbial metal reduction (Fig. 4.4).[31–34]

A number of studies have also shown that anthraquinone-2,6-disulfonate (AQDS), a surrogate for quinone moieties in humic substances, can act as an effective electron-shuttling compound in Fe(III) reduction.[35–37] The reduced product, anthrahydroquinone-2,6-disulfonate (AH_2QDS) can reduce insoluble Fe(III) with accompanying regeneration of AQDS. AQDS can be reduced by most Fe(III)- and humic substances-reducing bacteria,[38] resulting in enhancement of microbial reduction of various Fe(III) minerals.[35–37]

As described above, reduction of insoluble Fe(III) and subsequent dissolution of Fe(II) can result in release of associated As (Fig. 4.1). Therefore, the presence of an extracellular electron-shuttling compound might also affect microbial As mobilization. A recent study has reported that addition of AQDS significantly enhanced reductive dissolution of both Fe and As from As(V)-laden Fe(III) precipitates in the presence of a DARB capable of reducing AQDS.[21] Aqueous As levels, however, were nearly unchanged in the same experiments using Al precipitates, indicating that AH_2QDS cannot catalyze the reduction of solid-phase As(V), although it has been reported to reduce other metals and radionuclides such as Cr(VI) and U(VI).[39,40] Thus, an electron-shuttling compound, AQDS, can indirectly enhance DARB-mediated As mobilization, via acceleration of Fe(III) reduction. Since As mobilization hardly occurred in the absence of microbial As(V) reduction, even though the microbial Fe(III) and AQDS reduction actively occurred, DARB could be vital for As removal from contaminated soil.

4.5 As Removal from Contaminated Soils by Use of DARB and AQDS

To investigate whether enhanced As mobilization as mediated by DARB and an electron-shuttling compound is suitable for As remediation, laboratory-scale experiments using two types of As-contaminated soils (soil L, 250 mg As/kg; soil H, 2400 mg As/kg) collected at an industrial site were conducted with a well-characterized DARB, *Bacillus selenatarsenatis*, and AQDS.[21] Soil slurries containing liquid culture were incubated with the DARB and/or AQDS and the concentration of aqueous As and Fe were determined after 7 d incubation (Table 4.1). Although the DARB effectively dissolved As from both soils, the addition of AQDS evidently enhanced the microbial As mobilization. In contrast, no remarkable dissolution occurred in the abiotic control experiments, regardless of the presence of AQDS. When incubated with both the DARB and AQDS, nearly 56% and 43% of initial As was removed to the aqueous phase from soils L and H, respectively. Consequently, the residual As in soil L was decreased to below the soil concentration standard set by the Soil Contamination Countermeasures Law in Japan (150 mg/kg), although higher levels of As still remained in

Table 4.1. Reductive dissolution of As and Fe from contaminated soils. The results show dissolved total As (T-As), Fe (T-Fe), and Fe(II) concentrations in soil L and H slurries containing liquid cultures after 7 d incubation.

	Soil L (mg/L)			Soil H (mg/L)		
	T-As	T-Fe	Fe(II)	T-As	T-Fe	Fe(II)
DARB + AQDS	7.1	25.6	25.4	50.7	7.3	7.6
DARB	4.1	6.1	5.0	24.9	1.6	2.1
AQDS	1.4	0.2	0.5	11.4	0	0.5
Control	2.0	1.8	1.5	10.3	0	0.4

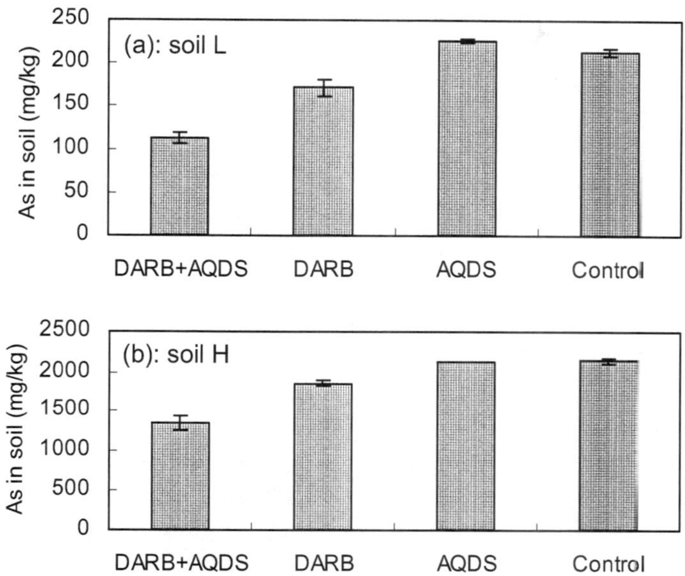

Figure 4.5. Residual As concentrations in contaminated soils L (a) and H (b) after 7 d incubation. Values represent the averages of two separate experiments, and error bars show the range.

soil H even when DARB and AQDS were both present (Fig. 4.5). Dissolved Fe, detected almost entirely as Fe(II), also increased with the addition of AQDS (Table 4.1), supporting the hypothesis that microbially produced AH_2QDS can indirectly accelerate As mobilization mediated by DARB, even in actual contaminated soils.

4.6 Conclusions and Outlook

This chapter has focused on the use of microbial As mobilization, especially solid-phase As(V) reduction mediated by DARB, as a potential mechanism for As removal from a solid phase. As accumulated in contaminated soils can be extensively removed to the aqueous phase by DARB, and an electron-shuttling compound can effectively improve the removal efficiency. This can be used as a bioremediation technique for contaminated soil containing As at the several hundred ppm level. Optimization of the removal conditions, the use of redox active vitamins as alternative electron-shuttling compound for AQDS[41], and combined application with chemical washing, such as with phosphate,[42] may lead to further cost effectiveness and/or adaptation of the microbial process to soils containing higher levels of As.

To establish a substantive remediation process using microbial As mobilization, further pilot-scale studies are imperative. The most practical system would be on-site treatment using a bioreactor to prevent diffusion of mobilized As as As(III) to downstream areas.[43] However, if a contaminated site has a low permeability zone below the contaminant plume, *in situ* treatment, using injection of microbially available carbon sources and electron-shuttling compounds and pumping up of the solution containing mobilized As(III) by separate wells, might also be conducted as a more cost-efficient biostimulation technique, because As(V)-reducing bacteria are widely distributed in the soil environment.[25] Although a subsequent treatment after solid/liquid separation in both systems is required to remove As(III) from solution, precipitation with sulfides and adsorption to Fe(II)-based solids is available for aqueous As(III) removal.[44–46] Thus, there is great potential for As bioremediation.

Acknowledgments

This work was supported by Grants-in-Aid for Young Scientists (B) (Nos. 18710069 and 20710063) from the Ministry of Education, Culture, Sports, Science and Technology of Japan.

References

1. Cullen, W. R. and Reimer, K. J., *Chem. Rev.* **89**, 713 (1989).
2. Smith, E., Naidu, R., and Alston, A. M., *Adv. Agron.* **64**, 149 (1998).
3. Belluck, D. A., Benjamin, S. L., Baveye, P., Sampson, J., and Johnson, B., *Int. J. Toxicol.* **22**, 109 (2003).

4. Morton, W. E. and Dunnette, D. A., in: *Arsenic in the Environment Part II: Human Health and Ecosystem Effects*, ed. J. O. Nriagu (Wiley-Interscience, New York), p. 17 (1994).
5. Davey, J. C., Nomikos, A. P., Wungjiranirun, M., Sherman, J. R., Ingram, L., Batki, C., Lariviere, J. P., and Hamilton, J. W., *Environ. Health Perspect.* **116**, 165 (2008).
6. Watson, W. H. and Yager, J. D., *Toxicol. Sci.* **98**, 1 (2007).
7. Yu, W. H., Harvey, C. M., and Harvey, C. F., *Water Resour. Res.* **39**, 1 (2003).
8. Mandal, B. K. and Suzuki, K. T., *Talanta* **58**, 201 (2002).
9. Mulligan, C. N., Yong, R. N., and Gibbs, B. F., *Eng. Geol.* **60**, 193 (2001).
10. U.S. Environmental Protection Agency, *Arsenic Treatment Technologies for Soil, Waste and Water* (2002).
11. Huang, Y.-C., in: *Arsenic in the Environment Part I: Cycling and Characterization*, ed. J. O. Nriagu (Wiley-Interscience, New York), p. 17 (1994).
12. Bhumbla, D. K. and Keefer, R. F., in: *Arsenic in the Environment Part I: Cycling and Characterization*, ed. J. O. Nriagu (Wiley-Interscience, New York), p. 51 (1994).
13. Dixit, S. and Hering, J. G., *Environ. Sci. Technol.* **37**, 4128 (2003).
14. Mok, M. W. and Wai, C. M., in *Arsenic in the Environment Part I: Cycling and Characterization*, ed. J. O. Nriagu (Wiley-Interscience, New York), p. 99 (1994).
15. Oremland, R. S. and Stolz, J. F., *Science*, **300**, 939 (2003).
16. Oremland, R. S. and Stolz, J. F., *Trends Microbiol.* **13**, 45 (2005).
17. Ahmann, D., Krumholz, L. R., Hemond, H. F., and Lovley, D. R., *Environ. Sci. Technol.* **31**, 2923 (1997).
18. Zobrist, J., Dowdle, P. R., Davis, J. A., and Oremland, R. S., *Environ. Sci. Technol.* **34**, 4747 (2000).
19. Yamamura, S., Yamamoto, N., Ike, M., and Fujita, M., *J. Biosci. Bioeng.* **100**, 219 (2005).
20. Cummings, D. E., Caccavo, F., Fendorf, S., and Rosenzweig, R. F., *Environ. Sci. Technol.* **33**, 723 (1999).
21. Yamamura, S., Watanabe, M., Kanzaki, M., Soda, S., and Ike, M., *Environ. Sci. Technol.* **42**, 6154 (2008).
22. Islam, F. S., Pederick, R. L., Gault, A. G., Adams, L. K., Polya, D. A., Charnock, J. M., and Lloyd, J. R., *Appl. Environ. Microbiol.* **71**, 8642 (2005).
23. Kocar, B. D., Herbel, M. J., Tufano, K. J., and Fendorf, S., *Environ. Sci. Technol.* **40**, 6715 (2006).

24. Tufano, K. J., Reyes, C., Saltikov, C. W., and Fendorf, S., *Environ. Sci. Technol.* **42**, 8283 (2008).
25. Yamamura, S., Watanabe, M., Yamamoto, N., Sei, K., and Ike, M., *Chemosphere*, **77**, 169 (2009).
26. Silver, S. and Phung, L. T., *Appl. Environ. Microbiol.* **71**, 599 (2005).
27. Langer, H. W. and Inskeep, W. P., *Environ. Sci. Technol.* **34**, 3131 (2000).
28. Bissen, M. and Frimmet, H. F., *Fresenius J. Anal. Chem.* **367**, 51 (2000).
29. Gaucia-Manyes, S., Jiménez, G., Padró, A., Rubio, R., and Rauret, G., *Talanta* **58**, 97 (2002).
30. Cancès, B., Juillot, F., Morin, G., Laperche, V., Alvarez, L., Proux, O., Hazemann, J.-L., Brown Jr., G. E., and Calas, G., *Environ. Sci. Technol.* **39**, 9398 (2005).
31. Lovley, D. R., Coates, J. D., Blunt-Harris, E. L., Phillips, E. J. P., and Woodward, J. C., *Nature*, **382**, 445 (1996).
32. Straub, K. L., Kappler A., and Schink, B., *Methods Enzymol.* **397**, 58 (2005).
33. Lovley, D. R., Fraga, J. L., Blunt-Harris, E. L., Hayes, L. A., Phillips, E. J. P., and Coates, J. D., *Acta Hydrochim. Hydrobiol.* **26**, 152 (1998).
34. Kappler, A., Benz, M., Schink, B., and Brune, A., *FEMS Microbiol. Ecol.* **47**, 85 (2004).
35. Hacherl, E. L., Kosson, D. S., Young, L. Y., and Cowan, R. K., *Environ. Sci. Technol.* **35**, 4886 (2001).
36. Royer, R. A., Burgos, W. D., Fisher, A. S., Jeon, B.-H., Unz, R. F., and Dempsey, B. A., *Environ. Sci. Technol.* **36**, 2897 (2002).
37. Dong, H., Kukkadapu, R. K., Fredrickson, J. M., Zachara, J. M., Kennedy D. W., and Kostandarithes, H. M., *Environ. Sci. Technol.* **37**, 1268 (2003).
38. Coates, J. D., Ellis, D. J., Blunt-Harris, E. L., Gaw, C. V., Roden, E. E., and Lovley, D. R., *Appl. Environ. Microbiol.* **64**, 1504 (1998).
39. Fredrickson, J. K., Kostandarithes, H. M., Li, S. W., Plymale, A. E., and Daly, M. J., *Appl. Environ. Microbiol.* **66**, 2006 (2000).
40. Jeon, B.-H., Kelly, S. D., Kemner, K. M., Barnett, M. O., Burgos, W. D., Dempsey, B. A., and Roden, E. E., *Environ. Sci. Technol.* **38**, 5649 (2004).
41. Yamamura, S., Watanabe, M., Soda, S., and Ike, M., *Book of Abstracts; XIX International Symposium on Environmental Biogeochemistry*, 58 (2009).
42. Alam, M. G. M., Tokunaga, S., and Maekawa, T., *Chemosphere*, **43**, 1035 (2001).

43. Soda, S., Kanzaki, M., Yamamura, S., Kashiwa, M., Fujita, M., and Ike, M., *J. Biosci. Bioeng.* **107**, 130 (2009).

44. Rittle, K. A., Drever, J. I., and Colberg, P. J. S., *Geomicrobiol. J.* **13**, 1 (1995).

45. Newman, D. K., Beveridge, T. J., and Morel, F. M. M., *Appl. Environ. Microbiol.* **63**, 2022 (1997).

46. Roberts, L. C., Hug, S. J., Ruettimann, T., Billah, M. M., Khan, A. W., and Rahman, M. T., *Environ. Sci. Technol.* **38**, 307 (2004).

Chapter 5

PHYTOREMEDIATION OF BORON POLLUTED SOIL AND WATER

Kazuhiro Mori[1] **and Motoyuki Asada**[2]

[1]*Department of Civil and Environmental Engineering, Interdisciplinary Graduate School of Medicine and Engineering, University of YAMANASHI/4-3-11 Takeda, Kofu-city, Yamanashi 400-8511, Japan*

[2]*Institute of technology, Shimizu corporation/3-4-17 Ecchujima, Koto-ku, Tokyo 135-8530, Japan*

mori@yamanashi.ac.jp; asada@shimz.co.jp

5.1 Introduction

Phytoremediation is a plants-using environmental cleaning technique applicable to organic and inorganic pollutants in the soil and water. The typical functions of the plants in the phytoremediation system are summarized to the absorption/adsorption and accumulation of inorganic matters. And we also have to pay attention to one more important mechanism of phytoremediation, the biodegradation or bioconversion of pollutants by rhizo-microorganisms. The water purification activities of more than 60 kinds of water plants against eutrophication have been studied from 1960s.[1] For the soil purification, phytoremediation of organic and heavy metal

Handbook of Metal Biotechnology: Applications for Environmental Conservation and Sustainability
Edited by Michihiko Ike, Mitsuo Yamashita, and Satoshi Soda

www.panstanford.com

contaminants have been mainly studied.[2–5] But we cannot find out any study on application of phytoremediation techniques to the boron polluted site. Soil quality criteria of boron in Japan (dissolved amount by elution test) is 1.0 mg/L, and its content criteria is 4000 mg/L. Generally, boron content in natural soil is below 100 mg/kg. This element exist as borate and widely used for raw materials of enamel, glass, or insecticide. In 2006, 4.7% of soil contaminations exceeding the soil criteria were boron contamination in Japan. Boron is a bio-essential element and accumulated in the terrestrial tissue including leaves in the plants. Recent studies[6,7] indicate that boron form the rhamnogalacturonan II-borate dimer in pectin, and this structure contributes to the cell wall stability. These fundamental characters of plant boron usage show the possibility of boron phytoremediation by the plant uptake, accumulation and removal of the contaminants with biomass recovery.

In this section, our research data of boron removal activities of higher plants are summarized.

5.2 Boron Uptake Capabilities

For the hydroponic culture tests, giant duckweed (*Spirodela polyrhiza*), common reed (*Phragmites australis*), manchurian wild rice (*Zizania latifolia*), common cattail (*Typha latifolia*), umbrella plant (*Cyperus alternifolius L.*), water spinach (*Ipomoea aquatic*), and Purple loosestrife (*Lythrum salicaria*) were purchased from nursery companies and inoculated into cultivation vessels with Hutner solution[8] as a basal medium and glass beads instead of soil for emergent plants in a green house. The average atmospheric temperature was 29.9°C, and light intensity was 6000 lux in the green house. Lighting system was operated with a daily illuminating program of 16 hours of light and 8 hours dark condition. For the soil cultivation tests, komatsuna (*Brassica rapa var. peruviridis*) were sowed in the soils supplemented with boron.

Three material emergent plants or 50 duckweeds were inoculated into the test vessel with various concentrations (0, 1, 5, 10 mg L^{-1}) of boron including basal media. After inoculation, boron

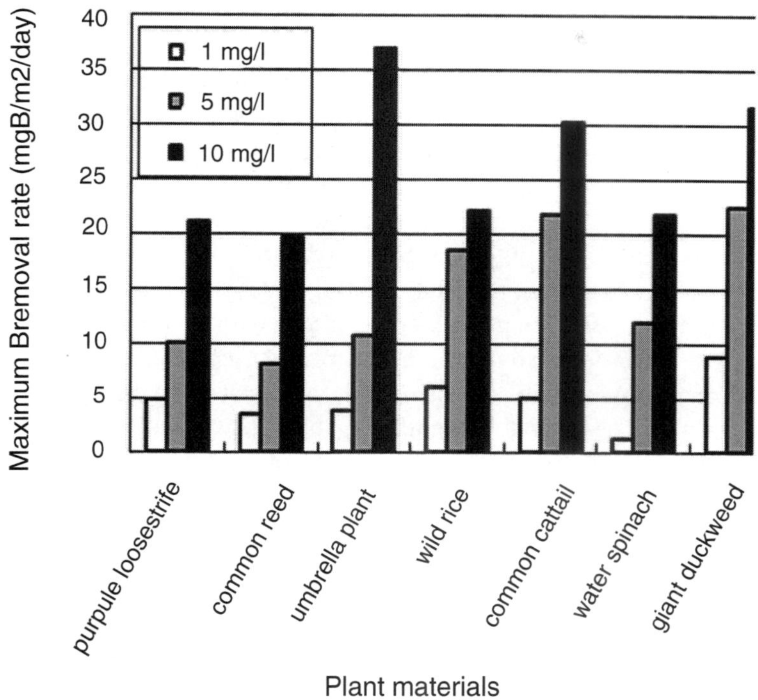

Figure 5.1. Maximum boron removal rate observed in the hydroponic cultivation tests.

concentrations of culture solution were analyzed every two days. The boron contents in the final plant tissues were also measured after boron removal tests. Boron concentrations in the solution were measured according to JISK0102. Boron extraction from the tissue and analysis were also operated according to the method reported by Matoh *et al.*[9]

We measured the boron absorption by several kinds of water plant. Figure 5.1 shows the maximum boron removal rate observed in the hydroponic culture of material plants. The removal rate depended on the boron concentration of the solution, and these relationships could be expressed as a following first-order equation (coefficient of determination: 0.87–0.98).

$$-\frac{dC}{dt} = k \cdot C, \tag{5.1}$$

where c is boron concentration (mg L^{-1}); t, time (h); and k, removal rate constant (h^{-1}).

These results show a possibility of estimate of boron uptake capability by plants, and also of phytoremediation of boron contaminated soil and water as a controllable engineering technique. We have constructed a model[10] which can estimate the biomass production by inputting environmental data sets of temperature, light intensity, and nutrients concentrations applicable to wide range of plants species including floating, emerged, and submerged water plants. Combining this biomass model with a model for boron content in the plant biomass, boron removal capabilities of each of plants can be simulated using data of climate and pollution of target site. These model techniques make plant selection, system planning, and designing in the manner of engineering possible. Such engineering support system would be applicable to another inorganic component after construction of database system for parameters of each of plant species. In this manner, phytoremediation system could be developed as a remediation technique on the engineering base, while this technique has been operated on experimental knowledge of experts. These recent results of our research suggest the fair chance of phytoremediation for boron contaminated soil and water. Some plant materials, however, showed growth inhibition under high boron concentration. We have to pay attention to the tolerance to the contaminants in the selection of plant materials for phytoremediation. And more, translocation of contaminants have to be taken in the consideration in case of operation of soil remediation applying the model system.

5.3 Boron Uptake and Removal from Soil by Plants

Twenty-five seeds of komatsuna (*Brassica rapa var. peruviridis*) were sowed in the soils supplemented with boron of 42 mg/kg to 4000 mg/kg (75 mg/kg of soil content is equivalent to 1.7 mg/L of quantity of elution test and 150 mg/kg of soil content is equivalent to 6.6 mg/L of quantity in the elution test). Water contents of the soil were maintained around 50–60%. The plantlets of komatsuna were recovered and the germination rate and biomass were measured two weeks after seeding.

5.3.1 *Effects of Boron on Komatsuna Plant After Germination in the Soil*

Germination rate is shown in Fig. 5.2. Almost 85% of seeds were germinated in the normal level (42 mg/kg) of boron content, and similar rate were observed up to 250 mg/kg boron content. On the other hand, only 18% of germination was found under the boron contamination of 500 mg/kg, and no seeds were geminated with over 1000 mg/kg contamination. These results show the material plant grow up under high boron contaminated situation comparing with environmental standards of Japan (1.0 mg/L as elution tests value). The leaf length and biomass weight are shown in Table 5.1. Growth of komatsuna was inhibited over 75 mg/kg boron content. These results indicate that phytoremediation can be applicable to the boron contaminated soil if the content of contaminant is 2–5 mg/L as elution tests value.

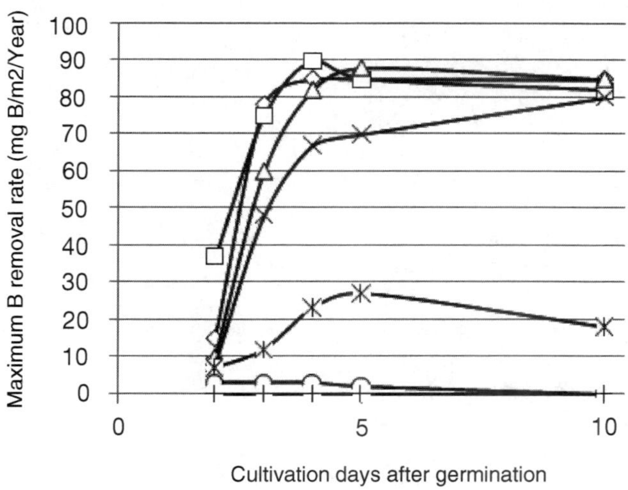

Figure 5.2. Influence of boron content in the soil on germination. □: 42 mg/kg, □: 75 mg/kg, □: 150 mg/kg, ×: 250 mg/kg, ∗: 500 mg/kg, ○: 1000 mg/kg, †: 4000 mg/kg.

Table 5.1. Influence of boron content in soil on leaf length and fresh weight of komatsuna.

B contents (mg/kg)	Leaf length (cm)		Fresh weight (after 20 days)	
(elution tests value; mg/L)	10 days	20 days	g/pot	Relative weight (%)
42	1.5	4.7	3.6	100
75 (1.7 mg/L)	1.5	4.5	3.7	104
150 (6.6 mg/L)	0.8	1.1	0.5	14
250–4000	0.7	—	—	—

5.4 Plants Detected in the Boron Contaminated Fields

Boron content in the tissue of plants growing in the boron contaminated fields (130–140 mg/kg according to the method in Soil Contamination Countermesures Law of Japan) was analyzed.

Table 5.2. Boron content in the plant tissue.

	B content (mg/kg)			
Plant	Root	Stem	Leaf	Flower
Woody plans, Dicotyledoneae class Archichlamiidae subclass: Japanese mulberry, Japanese hackberry, Muku tree, Japanese Mulberry, Locust tree	49–77	47–75	230–610	—
Harbaceous plants, Dicotyledoneae class Archichlamiidae subclass: American yellowrocket, Twotooth achyranthes, powleweed	61–130	71–130	310–1400	630
Woody plans, the Dicotyledoneae class Sympetalae subclass: Indian oleander, Japanese red elder	180	57–60	350–410	—
Harbaceous plants, Dicotyledoneae calss Sympetalae subclass: thorn apple, American black nightshade, horseweed, hogweed	54–86	56–76	160–990	—
Woody plans, Monocotyledoneae class: Chusan palm	79	83	140	—
Harbaceous plants, Monocotyledoneae class: wavyleaf basketgrass, Japanese lawngrass, Arundinaria chino, foxtail grass, Asiatic dayflower	110–240	41–230	70–600	67

Table 5.2 shows detected boron content in the tissues. Much amount of accumulated boron was detected in the leaf tissues comparing with root and stem. In the woody plans, Japanese mulberry, Japanese hackberry, and Muku tree under Dicotyledoneae class Archichlamiidae subclass, and Japanese red elder and Indian oleander under the Dicotyledoneae class Sympetalae subclass showed high boron content in the leaf tissues. In the herbaceous plants, poleweed (Dicotyledoneae class Archichlamiidae subclass), hogweed (Dicotyledoneae class Sympetalae subclass), and Asiatic dayflower (Monocotyledoneae) accumulated high content of boron in their leaves. The boron contents of pokeweed, hogweed, and Asiatic dayflower tissues were two to three times higher than soil, indicating a possibilities of soil remediation of boron polluted soil.

5.5 Conclusion

As mentioned above, plant uptake capabilities of boron can be applicable to the remediation of slightly polluted site. Plants maintenance techniques including model simulation and biomass usage have been developed as known well in these days. Strong points of phytoremediation are low energy consumption and easy maintenance against other remediation techniques. Higher removal activities, however, are requested because the operation term is longer than another method. Introduction of novel approach to accelerate the removal activities and to optimize the system would be the next future problem.

References

1. Fujita, M., Morimoto, W., Kono, H., Perdomo, S., Mori, K., Ike, M., Yamaguchi, K., and Soda, S., *Jpn. J. Environ. Sci.* **14**(1), 1–13 (2001) (in Japanese).
2. Krunger, E. L., Anderson, T. A., and Coals J. R., eds., *Phytoremediation of Soil and Water Contaminants* (American Chemical Society, WA) (1997).
3. Terry, N. and Banuelos G., eds., *Phytoremediation of Contaminated Soil and Water* (Lewis Publishers, Boca Raton, FL) (2000).

4. Fiorenza, S., Oubre, C. L., and Ward, C. H., *Phytoremediation of Hydrocarbon-Comtaminated Soil* (Lewis Publishers, Boca Raton, FL) (2000).
5. Mori, K. and Fujita, M., *Water Cleaning Using Plants, Environmental Water Cleaning Technology* (CMC Publishing, Tokyo), pp. 53–87 (2004) (in Japanese).
6. Hu, H. and Brown, P. H., *Plant Physiol.* **105**, 681–689 (1994).
7. Ishii, T., Matunaga, T., and Hayashi, N., *Plant Physiol.* **126**, 1698–1705 (2001).
8. Hutner, S. H., in: *Growth and Differentiation in Plants*, ed. Loomis, W. W. (Iowa State College Press, Ames), pp. 417–446 (1953).
9. Matoh, T., Akaike, R., and Kobayashi, M., *Plant Soil* **192**, 115–118 (1997).
10. Malla, R., Nagao, N., Maezono, K., Tanaka, Y., and Mori, K., *Proceeding of the 6th International Symosium on Southeast Asian Water Environment* 5–8 (2008).

Chapter 6

PHYTOREMEDIATION OF CADMIUM CONTAMINATED SOIL: ACCELERATION OF PHYTOREMEDIATION BY COMBINATION USE OF CHELATING AGENTS

Kazunari Sei

Division of Sustainable Energy and Environmental Engineering, Graduate School of Engineering, Osaka University, 2-1 Yamada-oka, Suita, Osaka 565-0871, Japan
Department of Health Science, School of Allied Health Sciences, Kitasato University, 1-15-1 Kitasato, Sagamihara-Minami, Kanagawa 252-0373, Japan
ksei@kitasato-u.ac.jp

6.1 Introduction

Cadmium (Cd) is a naturally occurring heavy metal in minerals and soils, and produced as a byproduct in zinc ore refinery process. World Cd production was 17,441 tonnes in 2007, and it was mainly produced by South Korea (20.1%), China (17.2%), Japan (11.1%), Mexico (9.1%), and Canada (8.0%) (People's Association on Countermeasures of Dioxin & Endocrine Disruptors, Japan, unpublished data). Cd is widely used for electrode in Ni-Cd battery, alloys, coating, pigment, and other industries. Among them, the use of Cd for Ni-Cd battery has been increasing up to more than 90%

Handbook of Metal Biotechnology: Applications for Environmental Conservation and Sustainability
Edited by Michihiko Ike, Mitsuo Yamashita, and Satoshi Soda

www.panstanford.com

of the total consumption (People's Association on Countermeasures of Dioxin & Endocrine Disruptors, Japan, unpublished data). Once Cd is ingested, it is slowly excreted, resulting in its quite long biological half-life (BHL). In case of human, BHL is estimated for 20 years.[1] In human body, Cd is mainly accumulated in liver and kidney. The highest accumulated concentration is observed especially in kidney. Cd accumulation in kidney exceeding an undisclosed level causes adverse effect to kidney function, especially renal tubular dysfunction such as reabsorption of calcium, phosphorus, and other minerals.[1] As a result, osteomalacia, well known in Minamata disease, and osteoporosis develop due to decrease of minerals in bone. Other chronic diseases such as emphysema and proteinuria can also be caused.[2] In addition, Cd and its compounds were classified as a human carcinogen (Group 1) by International Agency for Research on Cancer in 1993.

Cd is released into environment mainly by anthropogenic activities such as mining development and ore refining. In EU countries, strict Cd use restriction in electrical and electronic devices has come into effect from 2006 by Restriction of Hazardous Substances; however, the Cd applications that continue to grow are all centered around the Ni-Cd battery which has proved to be a very reliable, cost-effective battery for many applications in spite of the development of many other rechargeable battery chemistries and proposals for restrictions on Ni-Cd batteries in the EU. Thus, the increase in the Cd contamination cases remains possible. In Japan, Soil Contamination Countermeasures Act has been issued in 2003, and since then, soil contamination has been investigated in the sites for a factory or workplace for an abolished specified facility using hazardous substances, or in the land with suspected threat of health hazard by soil contamination. In 2003, 57 sites were revealed to be contaminated by Cd over standard level (Ministry of Environment, Japan).

Present treatment technologies for Cd contaminated soils are physicochemical methods such as containment and solidification/insolubilization by chemical addition. However, these methods generally require high cost to operate and give much environmental load, and what is more, the treated soils lose their function as a soil. On the other hand, in recent years, low-cost and environment-friendly soil remediation technology applying plants, phytoremediation, has been gathering attention. Use of hyperaccumulators, which can accumulate metals in their body at high concentration,[3] is considered to be one of the effective metal contaminated soil remediation. Generally, metals in soil are tightly adsorbed onto soil particles and soil organic matter. Therefore, for the effective phytoremediation by use of hyperaccumulators, it is essential to increase plant availability of such adsorbed metals. This strategy is called induced phytoremediation, and in the case of Cd, plant availability is increased by adding chelating agents resulting in eluting Cd tightly adsorbed in soil. However, this method has possibility to cause problems: (1) eluting more amount of Cd resulting in expanding the Cd pollution to surrounded groundwater and soil,[4] (2) eluting other minerals which are essential for plants growth resulting in deterioration of soil,[4] and (3) not always increasing the amount of Cd accumulated in plant body. Thus, the optimization on species and amount of the chelating agents is needed.

In this chapter, phytoremediation technology by the combination use of leaf mustard (*Brassica juncea*) and chelating agents for Cd contaminated soil remediation is introduced from our basic experiments on selection of chelating agents, optimization of their concentration, and their effect on the growth of *B. juncea* and Cd accumulation.

6.2 Biodegradability and Cd Elution Effect of Chelating Agents

The desired characteristics of chelating agents for soil remediation are that they are easily biodegradable in soil so as to avoid secondary pollution and that they can effectively elute target metals. Three representative Cd chelating agents — ethylenediaminetetraacetate (EDTA), ethylenediaminedisuccinate (EDDS), and citrate — are evaluated for their biodegradability and Cd eluting ability.

Each chelating agent was added at 0.1 mM (final concentration) to 100 mL basal salt medium inoculated with 1 g of garden soil and incubated at 28°C with rotary shaking at 120 rpm. By

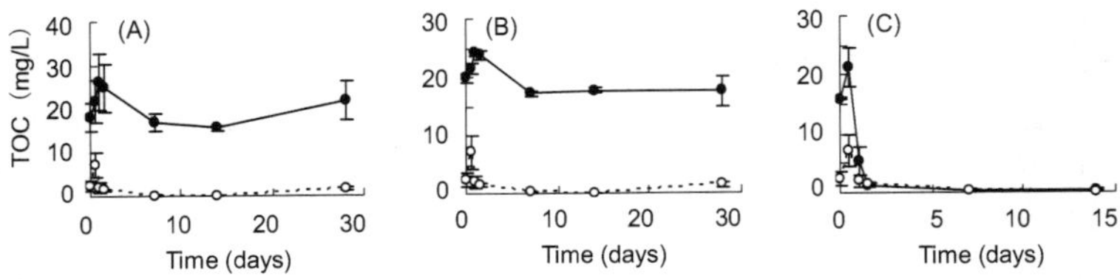

Figure 6.1. Biodegradability of three chelating agents, EDTA (a), EDDS (b), and citrate (c). TOC concentration was measured in the test systems with chelating agents (closed circle) and without chelating agents (open circle: control). Error bars represent standard deviation ($n = 3$).

periodically measuring total organic carbon (TOC) concentration, biodegradability was evaluated. As a result, citrate was completely degraded within one to two days, while EDTA and EDDS remained undegraded for more than 30 days (Fig. 6.1). This means that if EDTA and EDDS are used, they tended to persist in soil environment for long period.

Then artificial Cd contaminated soil was made using paddy soil by adding Cd at a final concentration of 0.1 mmol Cd/kg dry soil (112 mg Cd/kg dry soil), mixing thoroughly and incubating for seven days. Each chelating agent was added to be 0.1–40 mmol/kg dry soil to 10 g of sterile/non-sterile artificial Cd contaminated soil, ion-exchanged water was added to be 100 mL of total volume, and Cd elution ratio was periodically measured. Increased Cd elution was confirmed along with the concentration of chelating agents added, while no significant Cd elution was observed in the control (without chelating agents). The highest Cd elution ratio was confirmed in EDTA amendment system: 80% and 60% in sterile and non-sterile soils, respectively. EDDS gave a little lower elution ratio of 60% and 40% in sterile and non-sterile soils, respectively, and that of citrate was 30% and 40%, respectively. When citrate was added to non-sterile artificial Cd contaminated soil, the highest Cd elution ratio was observed on 1.5 days, and then gradually decreased to almost the same level with control after seven days (Fig. 6.2). In the case of EDDS, same as citrate, the highest Cd elution ratio was observed on two days, and then gradually decreased until seven days (Fig. 6.2). This is because the chelating agents added were biodegraded in soil, and resulted in eluted Cd was re-adsorbed onto soil. This suggests

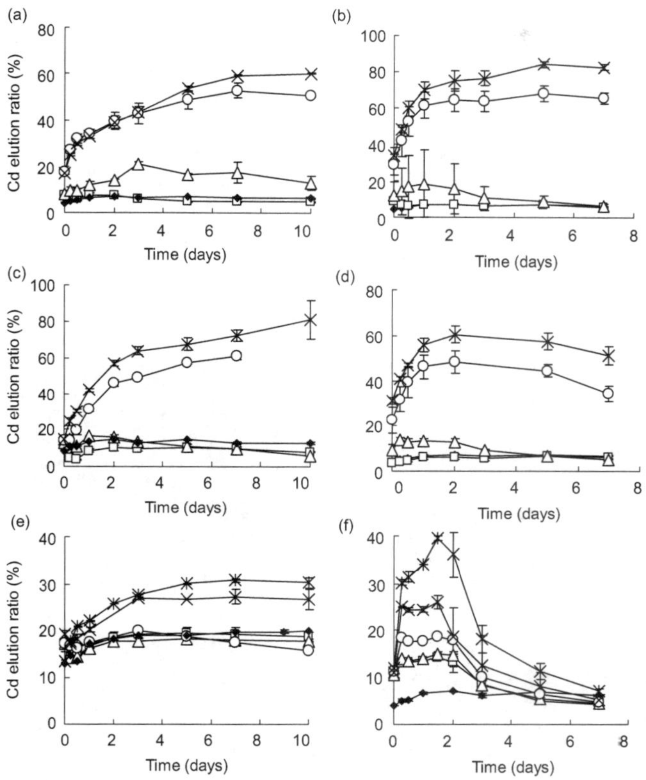

Figure 6.2. Time course of Cd elution ratio by EDTA (a and b), EDDS (c and d), and citrate (e and f) in sterile soil (a, c, and e) and non-sterile soil (b, d, and f) with concentration of 0 (closed diamond), 0.1 (open square), 1 (open triangle), 10 (open circle), 20 (cross), and 50 (asterisk) mmol/kg dry soil. Error bars represent the standard deviation ($n = 3$).

that it is important to consider the adding way of chelating agents along with the Cd absorption characteristics of the plants used when easily biodegradable chelating agents are applied.

6.3 Effect of Chelating Agents on the Growth of *B. juncea*

When chelating agents are applied for induced phytoremediation of Cd contaminated soil, it is important to evaluate the effect of chelating agents on the growth of plants. Selecting 50 days grown

B. juncea from seeding in the same size, each chelating agent was added at the concentration of 2 or 5 mM in hydroponic culture. *B. juncea* was further grown under the condition at 22°C and 10,000 lux (12-hour light/dark), and the growth of *B. juncea* was compared to control without adding chelating agents. Effect of chelating agents on water content of plant body was also evaluated. EDTA and EDDS clearly inhibited the growth of *B. juncea*, while citrate did not affect the growth of *B. juncea* (Fig. 6.3). EDTA and EDDS also lowered the water content of the plant (Fig. 6.4). Because Cd is absorbed from

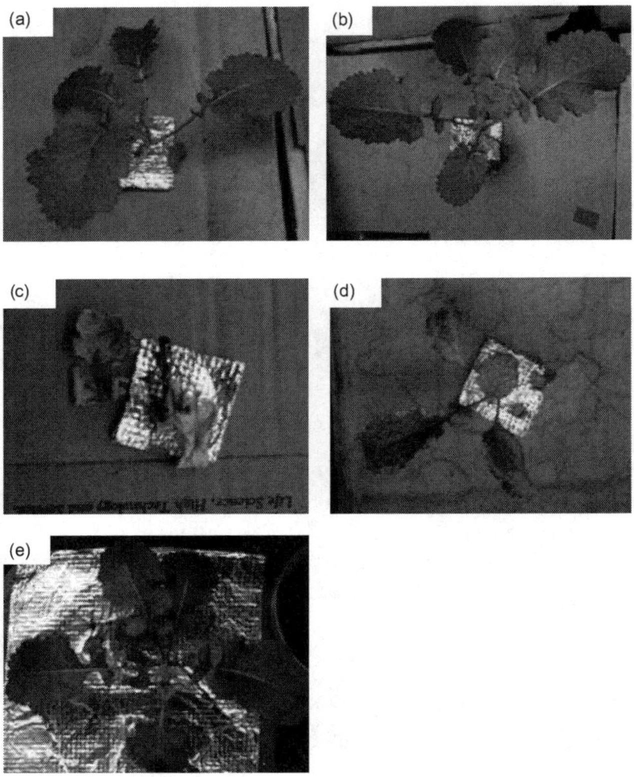

Figure 6.3. Effect of chelating agents on the growth of *Brassica juncea*. Photographs of *B. juncea* at the beginning of the experiment (a), after seven days without chelating agents (b), after seven days with EDTA (5 mM) (c), after seven days with EDDS (5 mM) (d), and after seven days with citrate (5 mM) (e) are shown. See also Color Insert.

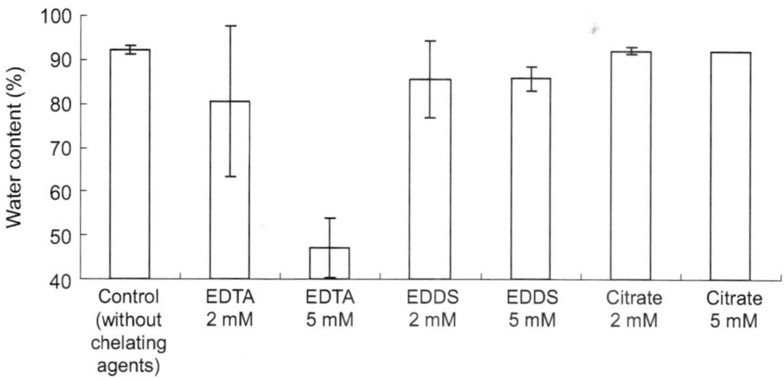

Figure 6.4. Effect of the chelating agents on water content of *Brassica juncea*. Error bars represent standard deviation ($n = 3$).

root along with water, EDTA and EDDS have a possibility to inhibit Cd absorption by *B. juncea* even if they effectively eluted Cd from soil.

Thus, for the successfully induced phytoremediation, it is important to evaluate the effect of chelating agents on plant physiology in addition to the eluting ability of target metals and biodegradability in soil.

6.4 Effect of Chelating Agents on the Accumulation of Cd in *B. juncea*

To evaluate how the Cd accumulating property of *B. juncea*, the Cd hyperaccumulator, changes, each chelating agent was added at 0.2–40 mM to hydroponic culture of *B. juncea* with 0.2 mmol Cd/L (22.4 mg Cd/L), and accumulated Cd was measured with its distribution in aerial and underground parts. Maximum Cd accumulation was observed when the chelating agents were added at 0.2 mM in the case of EDTA and EDDS, and at 40 mM in the case of citrate; however, total amount of Cd accumulated in *B. juncea* decreased (Fig. 6.5). This suggests that Cd ion form (Cd^{2+}) is easier to be absorbed by *B. juncea* than the chelated Cd. From the viewpoint of Cd distribution in the *B. juncea* body, 16% of the accumulated Cd was in aerial part in the control (without chelating agents). On the other hand, 47–98%

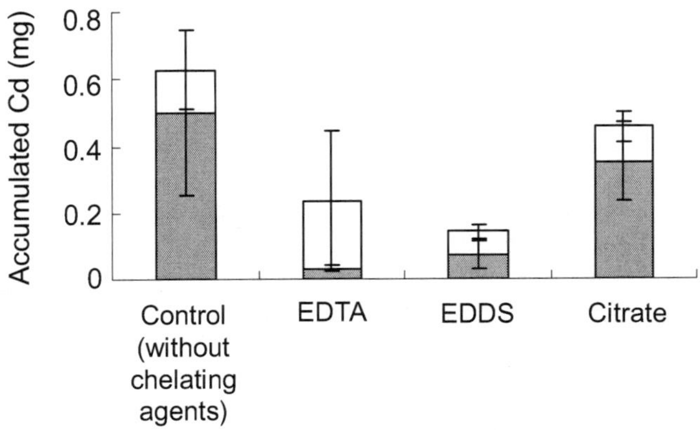

Figure 6.5. Effect of the chelating agents on the amount of Cd accumulation in *Brassica juncea* (hydroponic cultivation) and the distribution of accumulated Cd in aerial part (open bar) and in underground part (closed bar). Error bars represent standard deviation ($n = 3$).

and 50% of the accumulated Cd was confirmed in aerial part when EDTA and EDDS were added, respectively (Fig. 6.5). Thus, chelating agents enhanced the Cd transport to aerial part in the *B. juncea* body. Considering the practical application for phytoremediation, it is not enough by letting plants absorb Cd but the plant which accumulated Cd should be harvested. From this viewpoint, if the more Cd accumulated in aerial part than in underground part, the removal of Cd from soil become easier because the aerial part of the plant is the main part to be removed from the contaminated site. This is another advantage of phytoremediation by combination use of chelating agents.

6.5 Feasibility Study on Phytoremediation of Cd Contaminated Soil by Combination Use of *B. juncea* and Chelating Agents

The effect of the addition of chelating agents on Cd accumulation in *B. juncea* was evaluated using artificial Cd contaminated soil. The Cd contaminated soil was prepared at 0.1 mmol/kg dry soil (11.2

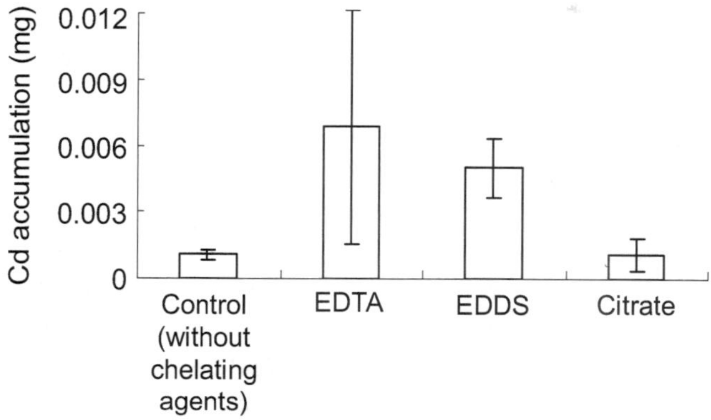

Figure 6.6. Effect of the chelating agents on the accumulation of Cd in *Brassica juncea* (soil experiment). Error bars represent standard deviation ($n = 3$).

mg Cd/kg dry soil). The 50 days grown *B. juncea* was planted in the Cd contaminated soil with adding each chelating agent at 2–20 mmol/kg dry soil, and 10 days Cd accumulation test was performed. As a result, in the case of EDTA and EDDS, the highest Cd accumulation was confirmed at 5 mmol/kg dry soil, and in the case of citrate, the highest Cd accumulation was confirmed at 20 mmol/kg dry soil (Fig. 6.6). Especially when EDTA was used, the accumulated Cd amount was 6.5 times higher than the control (without chelating agents). In the case of the citrate, as the highest Cd accumulation was confirmed at the highest citrate concentration tested (i.e., 20 mmol/kg dry soil), the more accumulation of Cd might be possible if higher concentration of citrate was applied.

6.6 Future Perspective for the Effective Induced Phytoremediation of Cd Contaminated Soil

In this chapter, as a basic information to evaluate whether the phytoremediation by the combination use of *B. juncea* and chelating agents can be effective technology for Cd contaminated soil remediation or not, three representative chelating agents, EDTA,

EDDS, and citrate, were evaluated from the viewpoints of Cd eluting ability, biodegradability, the effects on plant growth, and Cd accumulation. EDTA has high ability to elute Cd and to accumulate Cd in *B. juncea* body with low additive amount. EDTA is also superior in promoting the accumulation of Cd in the aerial part of the *B. juncea* body. Thus, EDTA seems promising chelating agent for induced phytoremediation. However, at the same time, EDTA is less biodegradable, persists in soil for long period, and as a result, high Cd eluting effect remains for long period. This further leads the possible pollution expansion in uncontaminated surroundings. Therefore, it is important to optimize the additive amount and to manage the land after the treatment. EDDS has the second best ability to elute Cd and to accumulate Cd in *B. juncea* body. A certain biodegradation, which caused re-adsorption of eluted Cd to soil, was confirmed. Thus, the use of EDDS seems less risky to expand the Cd contamination to the surroundings than EDTA. On the contrary, citrate has low ability to elute Cd. To enhance the accumulation of Cd in *B. juncea* body using citrate, higher additive amount is needed than EDTA and EDDS. However, the biodegradability of citrate is high, and the Cd eluting ability will not remain longer. This means the addition of citrate will have less effect on natural soil property.

Thus, the three chelating agents evaluated here have their own advantages and disadvantages. The deciding factor in choosing the chelating agents is that on which property of the chelating agents the user sets much store. For example, if the target area is apart from the range of daily human activity, and no limitation in use of the land after remediation, EDTA with rapid, low-cost, and high Cd removal efficiency seems the best way. On the other hand, if the target area is paddy field, where it is desirable that the addition of chelating agents have less effect on the soil property, citrate seems a better choice. When adding chelating agents, it is essential to investigate and keep the adequate additive amount. Application of too much amount of chelating agents can not only cause the reduction in Cd removal efficiency but can also result in expanding the Cd contamination by eluting much amount of Cd beyond the ability of plant to absorb, the deterioration of soil by eluting other minerals essential for plant growth, and the secondary pollution by chelating agents. For the implementation of the practical induced phytoremediation, it is desirable to accumulate more information on the effect of chelating agents with expanding the species of chelating agent and plant, and build up the database for the selection of suitable chelating agent, plant, and operational method for each purpose to achieve the cleanup level.

References

1. Nordberg, M., *BioMetals* **17**, 589–592 (2004).
2. Bonnell, J. A., *Br. J. Ind. Med.* **12**, 181–197 (1955).
3. Brooks, R. R., *Plants That Hyperaccumulate Heavy Metals* (NY CAB International, USA) (1988).
4. Wu, L. H., Luo, Y. M., Xing, R. R., and Christie, P., *Agric. Ecosyst. Environ.* **102**, 307–318 (2004).

Chapter 7

PHYTOREMEDIATION FOR SOILS CONTAMINATED WITH HEAVY METALS USING THE SYMBIOSIS BETWEEN *ASTRAGALUS SINICUS* WITH RHIZOBIA

Mitsuo Yamashita
Department of Applied Chemistry, Faculty of Engineering, Shibaura Institute of Technology, 3-7-5 Toyosu, Koto-ku, Tokyo 135-8548, Japan
yamashi@sic.shibaura-it.ac.jp

7.1 Introduction

Bioremediation has evolved from management and treatment of municipal and industrial wastewater and solid wastes. Land disposal of wastewaters on sewage farms began in the late 19th century and involved the use of soil bacteria. In decontamination processes more sophisticated methods of contaminant treatment, such as tricking filters, activated sludge, and anaerobic fermentation, were advanced in the first half of the 20th century. Since 1960, biological treatment processes have continued to include new methods of land treatment and processes for biodegradation of particular types of compounds.

Handbook of Metal Biotechnology: Applications for Environmental Conservation and Sustainability
Edited by Michihiko Ike, Mitsuo Yamashita, and Satoshi Soda

www.panstanford.com

Developments in wastewater and solid-waste treatment have been transferred to the treatment of contaminated soils and groundwater. In recent years, most of the works on bioremediation has been on the treatment of soils contaminated with petroleum products, since most petroleum hydrocarbons are relatively easy to degrade and are amenable to bioremediation, and the large number of sites contaminated with petroleum hydrocarbons from leaking underground storage tanks.

Bioremediation has become increasingly important in the field of hazardous-waste management. Some of the chemicals that were once thought to be resistant to degradation, including chlorinated species such as trichloroethylene (TCE), and certain polychlorinated biphenyls, have been shown to be biodegradable at least under laboratory conditions. Other compounds that are currently being targeted by bioremediation include (1) solvents such as acetone and alcohols; (2) aromatics such as benzene, toluene, ethylbenzene, and xylenes, which are collectively known as BTEX, as well as polycyclic aromatic hydrocarbons, and chlorobenzene; (3) nitro- and chloro-phenols; and (4) pesticides. Aromatic hydrocarbons such as BTEX from spillage or leakage and chlorinated aliphatics such as tetrachloroethlene or perchloroethylene, TCE, and 1,1,1-tricholoroethane, used in industry for degreasing, are targeted among the most commonly encountered contaminants in soil and groundwater.

The success of bioremediation is often measured by the percent reduction in contaminant concentration in the soil or groundwater. Such criteria are weak since bioremediation may fail to achieve the high percent removals to meet cleanup goals. At the same time, the contaminants may be transported out of the soil or water through other processes, such as volatilization, migration, or photooxidation. In such a case, the goals of bioremediation, which are to detoxify and immobilize the contaminants, would not necessarily be net. A successful bioremediation process should include controls to account for contaminant transport, such as a cover to collect volatile material or monitoring wells to detect contaminant migration. At the same time, proof that biodegradation has occurred needs to be obtained. That can be in the form of increased microbial activity, increased release of carbon dioxide, increased uptake of oxygen, or presence of metabolic products. Bioremediation can be applied to any system or process, in which biological methods are used to transform or immobilize contaminants in soil or groundwater. One result of this very broad definition is that little knowledge is gained from saying that bioremediation has been or will be applied in a particular case. The physical conditions involved in bioremediation processes can be classified as *in situ*, solid phase, or bioreactors. Treatment processes used in the bioremediation of contaminated soils, gases, and water are different from each other considerably. A brief description of approaches and processes is presented here.

7.2 Remediation Treatment

7.2.1 *Groundwater Bioremediation*

Methods used for groundwater bioremediation can be categorized as pump and treat or *in situ*. In pump and treat systems groundwater is pumped to surface, treated, and either used directly or returned to the aquifer. Surface treatment often involves aeration, the addition of nutrients, and in some cases seeding with microbial cultures capable of degrading contaminants known to be in the groundwater. The used processes are similar to what is involved in wastewater treatment, such as activated sludge and rotating contactors. In some cases the surface treatment is physical as it may involve adsorption onto activated carbon or air stripping. Pump and treat operations have generally not been successful in remediation groundwater back to drinking water standards. The *in situ* treatment occurs in the subsurface. The activity of the degrading microorganisms in an aquifer is stimulated by the introduction of oxygen and nutrients. Oxygen is often found to be the most rate-limiting factor in subsurface biodegradation and can be introduced in a number of ways such as air spurge or through the use of hydrogen peroxide. Nutrients are usually introduced in a water solution through infiltration galleries or injection wells. Subsurface conditions are not easily controlled. The problems associated with *in situ* treatment include the difficulty in delivery of oxygen or an alternate electron acceptor at satisfactory rates or to the necessary locations, difficulty in delivering microorganisms to contaminated zones, isolation of contaminants in noncontiguous interstitial openings, and inability to prevent migration of contaminants beyond the treatment zone.

7.2.2 *Soil Bioremediation*

Bioremediation of contaminated soils can be carried out *in situ*, or the material can be excavated and treated *on site* or at a separate treatment facility. *In situ* treatment methods include soil venting, in cases where volatile contaminants are dominant, involving semi-volatile and non-volatile contaminants. *Ex site* processes include land treatment (usually a form of land improvement), composting, and slurry-phase bioreactors. Both *in situ* and *ex situ* treatments often involve a combination of biological and non-biological processes and operations. For example, soil washing is often used to concentrate contaminated materials and reduce the volume that must be treated. In soil washing surfactants are used to remove absorbed materials from larger components such as rock and gravel. Washing usually has little effect on materials absorbed to soil organic material, silt, and clay, but the separation of large materials makes biological treatment considerably simpler.

7.2.2.1 *In situ* bioremediation

In situ soil bioremediation requires transporting oxygen and nutrients through the contaminated volume. In some cases the indigenous microbial population is unsatisfactory in terms of present species and an enriched microbial culture needs to be added. Even highly porous soils present relatively severe limitations on transport of liquids and particles. For this reason, addition of nutrients and microorganisms is difficult. In cases where the contaminants are volatile (e.g., a gasoline spill) soil venting and *ex site* gas treatment can be applied. In soil venting air is drawn through the polluted soil zone and the off-gases are treated. Some biodegradation of the contaminants may occur in the soil, but usually nutrient limitations exist and the amount of biodegradation is minimal. In the past, soil venting operations have been operated as physical treatment processes, where the main objective was to recover as much of the volatile contaminants as possible. The soil venting operation is biological. Biological venting is a form of *in situ* treatment applicable to less volatile contaminants in soil. Air or oxygen is drawn into the contaminated zone and nutrients are added by infiltration or injection. If successful, biological venting is a cost-effective alternative to excavation and treatment on the surface.

7.2.2.2 Land improvement

Land improvement involves aeration and mixing of contaminated soil by cultivating, addition of nutrients and microorganisms, and control of moisture content by periodic addition of water. In most cases contaminated soils are dug out and treated at a site where migration can be controlled by construction of leaching barriers. In some cases the contaminated soils are near enough to the surface to make the dig step unnecessary and the treatment is effectively *in situ*. Degradation processes in land improvement are principally biological. Photochemical oxidation may be significant in some cases. Emission of contaminants to the atmosphere through volatilization is often a limiting constraint on the application of land improvement.

7.2.2.3 Bioreactors

Bioreactors are slurry-phase operations, in which the contaminated soil is placed in a containment vessel. Enough water is added to allow continuous mixing. Oxygen can be added as required and off-gas controls are often used to prevent loss of volatile organic compounds through stripping. Off-gas controls include gas recycling, use of the off-gases in combustion processes, and potentially, microbial gas cleaning.

7.2.3 *Heavy Metal Contaminants and Remediation Technologies*

The metals found in greatest abundance are mined in soil in abundance. Iron (Fe), chromium (Cr), lead (Pb), zinc (Zn), copper (Cu), and nickel (Ni) present the six major targets of mining operations, with cadmium (Cd), silver (Ag), and mercury (Hg). Metals are directly or indirectly involved in microbial growth and metabolism. Some of their interactions with microbes are of biotechnological importance, being relevant to metal removal and recovery from mineral deposits and industrial effluents for industrial use or environmental bioremediation. Currently employed non-biological treatments used for removing contaminants from metal bearing wastewaters include membrane separation, liquid extraction, carbon adsorption, ion exchange, electrolytic treatment, precipitation, coagulation/flocculation, reduction, flotation, glassification, evaporation, and crystallization.

Soil provides mechanical support and nutrients for plant and microbial growth. Fertile soils are characterized by both the presence of nutrients and a physical structure amenable to living organisms. A broad range of microorganisms, which contain bacteria, actinomycetes, fungi, algae, and protozoa, is nearly always present in soil, although population densities vary widely. The surface of soil granules is the site of many of the biochemical reactions that take place in the cycling of organic matter, nitrogen, and other minerals in the weathering of rocks and in the nutrition of plants.

In Japan, more than half of soil contaminants are caused by heavy metals. The result cause that six elements, Cd, Pb, Cr, Ag, Hg, selenium (Se), are banned as noxious matter and banned as environment quality standard in soil environmental pollutant standard. These elementals morphology has changed in absorbance in living cell, such as ion, oxo acid, and organic compounds.

Cd, one of the most toxic metals to man and other forms of life, has no known biological function. Cd compounds enter the environment through a variety of industrial processes, such as mining, smelting, and electroplating, and to a less extent, from natural weathering. The effects form human exposure via food chains in Cd-polluted areas and the known lethality to metal in case of extreme exposure have focused attention on developing methods for its removal from polluted areas. Unlike organic substances which can be changed by biological and chemical processes into innocuous final products, metals can only be changed in valence or chelating state. They need to be immobilized, so that they cannot enter surface water, or be recovered. Due to the increasing value of some metals, as well as the greater awareness of the human and ecological effects of toxic metals released into the environment, studies of metals accumulation have been focused on their removal and recovery from solutions. Some microorganisms showed great promise for use in metal recovery because metal ions can be either passively absorbed by microbial biomass or intracellularly accumulated by active uptake into living cells. Studies on Cd biosorption by bacterial, fungal, or algal biomass have been reviewed. Biosorption treatment using microbial biomass could be competitive to conventional ones, such as ion exchange resins, in terms of effectiveness and cost. However, they generally lack the desired specificity. Many eukaryote carry genes that allow them to take up and intracellularly accumulate metals such as Cd, cobalt (Co), Zn, etc. All such systems are remarkable for their high affinity and specificity for metal ions. The specificity and affinity of active uptake systems would allow bacterial cells to selectively remove a desired metal ion from dilute solutions where the concentration of the metals is too low for passive sorption or other conventional methods. The molecular bases of some of these uptake systems are being worked out. Metal bioaccumulation and recovery is an area to which genetic engineering has not been applied. Given the fact that recovery of highly toxic metals, such as Cd, Hg, requires active concentration of the metal ion, this area could benefit from recombinant DNA technology.

7.2.4 *Cd Purification*

Cd is heavy metal at an atomic number 48. Cd occurs as a minor component in most Ze ores and is refined as a by-product of Zn production. Cd is used largely as pigment, in Ni-Cd batteries, and for corrosion resistant planting on steel. Cd has high toxicity and carcinogenicity. The associated health and environment are concerned. Cd results in the development of toxic symptoms, such as vomit, difficulty in breathing, pulmonary emphysema, and liver functional disorder. There have been notable instances of toxicity to Cd in contaminated food and water. Maximum of Cd in food was determined for safe consumption of food. 0.4 ppm of Cd in polished rice is banned by Codex Alimentarius since 2006.

One of purification treatments of Cd in the soils is soil dressing, which is to change contaminant soil to uncontaminated soil. The other is the disturbance of Cd eluting using sulfur or purification and removal of Cd using absorbance chemicals such as chelate. This physico-chemical unit process has drawback in high cost and the poor fertility of soils.

Bioremediation is one of purification measures using microorganisms and has merits available for a variety of contaminants, and a few of wasted energy and purification cost in generally. In years, phytoremediation has been developing to recover contaminant environment using plants. The plants that accumulate specially heavy metals are called hyper-accumulator. *Thlaspi arvense*, *Brassica juncea*, and *Athyrium yokoscense* are hyper-accumulator for Cd. Phytoremediation also has drawback in long-time in purification, and available for a limited contaminants, a limited range around roots, and soluble contaminants.[1] To overcome these hard problems, author introduces a remediation system for heavy metals using a novel plant–bacterial system, called symbiosis, which gives multiplier effect.

7.3 Plant–Microbe Interactions

The study of symbiosis in plant–microbe has been developing in the system of leguminous plant and rhizobia or trees and micorrhizal fungi. The root hairs of plants produce flavonoids. NodD protein in rhizobia recognizes specific flavonoids and rhizobia induce and express *nod* genes. As a result rhizobia produce derivatives of chitin-oligosaccharide, called Nod factors out of cells. Host plants recognize Nod factors and take in the rhizobia. The rhizobia grow in the root of plants and form infection thread and go into the root hair tissue and form bacteroid, which is a bunch of root nodule rhizobia in symbiosis. As a result the root cause cell differentiation and form nodule. The plants supply saccharide and organic acids to bacteria as an energy source and the bacteria obtain host.[2] Rhizobia grow in an aerobic condition, but keep live in anaerobic condition in nodule.

7.4 Development of Platform in Symbiosis Engineering

The study of symbiosis in plant–microbe, and the design and application of symbiosis, is called "symbiosis engineering."[3] The symbiosis in leguminous plant and rhizobia is used for the following advantages: Since rhizobia are Gram-negative bacteria like *Escherichia coli*, the same vector system used in *E. coli* is available in rhizobia. Since rhizobia express some *nif* operon genes in bacteroid, the results suggest that multiple genes would be expressed. The plants fix CO_2 in the air by photosynthesis and accumulate saccharide and protein in seeds and fruits. The plants are environment-friendly and get water and nourishment in soil and absorb soluble chemicals depending on the extension of roots. *Astragulus sinicus* (Chinese milk vetch, or renge-soh in Japanese) is a typical legume used as a green manure in rice field (Fig. 7.1). At present renge-soh has been used as few green manure and is substituted for artificial manure. The flower of renge-soh is known as a source of syrup on the market. Previously Rhizobium B3, which is a nitrogen-fixing bacterium, was isolated from the nodules of renge-soh in Hiroshima in Japan. Rhizobium of renge-soh between China and Japan was found to be of different type. Rhizobium of renge-soh in Japan is renamed as *Mesorhizobium huakuii* subsp. Rengei.[4] To express useful genes in *M. huakuii* subsp. *rengei* the host–vector system was developed by

Figure 7.1. *Astragulus sinicus* (Chinese milk vetch, or renge-soh in Japanese). See also Color Insert.

electroporation using a broad-host range vector DNA, pBBR122, and pKY230.[5]

7.5 Application of Symbiosis Engineering in Remediation

Metallothionein (MT) is a low-molecular-weight, cystein-rich protein that binds heavy metals such as Cd, Zn, Hg, and Cu. Since human MT gene (*hMT*) was successfully overexpressed in *E. coli*,[6] further research is planned to focus on the accumulation of heavy metals among contaminants. To increase the binding ability of heavy metal and to stabilize, the gene was designed for tetrameric human MT (MTL4) using protein engineering. The gene was successfully overexpressed in *E. coli* to generate functional oligomeric h-MT, which bound 28 g atoms of both Cd and Zn, respectively.[7]

M. huakuii B3 in renge-soh harbor nitrogen fixation and regulation genes (*nif, fix*), and nodulation forming synthesis genes (*nod, nol*) on their large plasmid DNA.[8] Since at least the *nif* and *nol* genes are expressed in nodule bacteroid under anaerobic condition, it is suggested that when the promoter of *nif* and *nol* genes was cloned upper target genes and target genes was expressed, a new useful function will be added.

The promoter of *nifH* and *nolB* gene was isolated and *MLT4* gene was cloned under the each promoter. The plasmid DNA carrying the recombinant gene was introduced into *M. huakuii* B3 in renge-soh. The nodule in renge-soh synthesized MTL4 in bacteroid containing the recombinant *M. huakuii* B3.[9] This is the first report that a mammalian gene was expressed in bacteroids in the nodules (Fig. 7.2). The content of Cd in nodules containing bacteroid carrying *MLT4* gene increased by about twofold as compared with nodules containing bacteroids non-carrying *MTL4*. Since we clarified *hMT* bind arsenic (As),[10] the results suggest to apply bioremediation in As using a novel symbiosis.

Phytochelatins (PCs) are naturally occurring metal binding short peptides and an attractive alternative to MTs since PCs offer the potential for enhanced affinity and selectivity for heavy metals. The structure of such peptides can be represented by (g-Glu-Cys)*n*-Gly, where *n* ranges from 2 to 11. *Arabidopsis thaliana* gene for PC

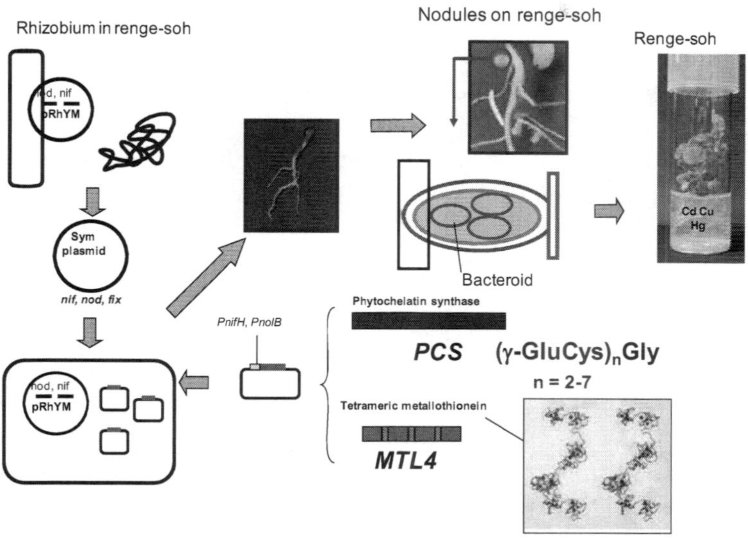

Figure 7.2. Creation of symbiosis biomaterial for metal purification. See also Color Insert.

synthase (*AtPCS*) was cloned and fused in frame downstream of *nifH* promoter. The expression of *AtPCS* gene was demonstrated in free-living cells of B3 under low-oxygen conditions. A range of PCs was synthesized by cells that expressed the *AtPCS* gene, with values of *n* from 2 to 7, whereas no PCs were found in control cells that harbored the empty plasmid. The expression of *AtPCS* gene in *M. huakuii* B3 increased the ability of cells to bind Cd about 9- to 19-fold.[11] The PCS protein was detected by immunostaining in bacteroids of mature nodules of renge-soh containing the *AtPCS* gene. When recombinant *M. huakuii* B3 established the symbiotic relationship with renge-soh, the symbionts increased Cd accumulation in nodules 1.5-fold. The *MTL4* and *AtPCS* genes were transferred to *M. huakuii* B3, which can infect and form nodules on renge-soh. When these recombinant strains established the symbiosis relationship with on renge-soh, the symbionts increased Cd accumulation in nodules by twofold in hydroponic culture.[15] The expression of the both *MTL4* and *AtPCS* genes showed additive effect on Cd accumulation in nodules. Authors also applied these recombinant bacteria to rice paddy soil polluted with Cd (1 mg kg^{-1} dry weight soil). The accumulation of

Cd in the plant roots infected by recombinant B3 achieved threefold than that by the wild-type B3. After two months of cultivation of the symbionts, a maximum of 9% of Cd in paddy soil was removed.[12] Thus, the symbiosis will be useful in phytoremediation for heavy metals.

To promote the transport of metals into the nodules of rhizobium and the accumulation of metals, the Fe-regulated transporter 1 gene from *A. thaliana* (*AtIRT1*) was cloned and fused in frame downstream of *nifH* and *nolB* promoter. The fused AtIRT1-alkaline phosphatase was expressed in the free-living recombinant B3 and the nodule of renge-soh. The strain B3:nifHPCS, which is inserted at *nifH* fused *AtPCS* gene in genome of B3, carrying *AtIRT1* expression vector, pMPnolBIRTHis, showed a 1.6-fold increase in amount of Cd accumulated compared with the strain B3:nifHPCS.[13] To examine where the expression of *AtIRT1* affects metal accumulation in bacteroids in nodules, the hydroponic culture tests using the plants with recombinant nodules were performed. No enhancement of Zn accumulation in the recombinant nodules was observed in the presence of AtIRT1 like in the case of Cd. However, the amount of Cu and As accumulated in nodules containing B3:nifHPCS carrying pMPnolBIRTHis was 1.3-fold higher than that in nodules containing B3:nifHPCS.[13] It is supposed that as an amount of metals accumulated in plants is increased, the value of resource is increase. After the plants were burned, the metals are recovered from the ash. The contaminant soil was purified by the repeated performance. The system is better and more safe than solidification and also more economical than soil dressing.

7.6 Development of Phytoremediation Technology for Metal Contaminants

To raise the purification efficiency for metal using the symbiosis in plant-microbe, I mention the ideas as follows (Fig. 7.3): (1) an increased amount of metals accumulated in plants by the increased production of biomass, (2) a promotion of absorbance in plants by solubilization of heavy metals in soils, (3) a creation of genetically

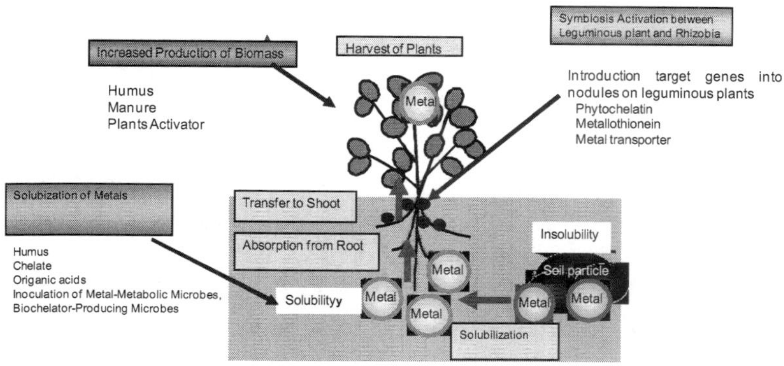

Figure 7.3. Strategy of phytoremediation technology for metal contaminants. See also Color Insert.

modified organization with genes coded metal absorbance, metal import, and metal binding proteins.

To increase production of biomass, humus and activators for plants were administrated. So, authors investigate to use a variety of chemicals as environmental purification. Chelate chemicals and organic acids (citrate, oxalate) were given generally to solubilize heavy metals in contaminant soils. However, chelate chemicals and organic acids were known poorly for their characterization of biodegradation and effects of elution. Authors examined the addition of fungi that overproduce organic acids and screening of microorganisms that overproduce of biodegradable chelate on the point of view from environmental pollution control. As a result, biomass is increased according to the production of organic acids and the amount of Cd in plants is increased for promotion of Cd solubility. The results suggest that the plants co-growth up with microorganisms except rhizobium and are able to purify contaminated soil and use for remediation.

7.7 Future of Metalbiotechnology in Symbiosis Technology

Authors substituted aspartic acid and serine in *hMT* gene for cysteine by protein engineering. The mutant hMT protein binds with more affinity for Cd than wild-type hMT.[14] Enlarged on the successful result, it may that we create the protein new metal binding sites, as the protein has never been bound for a new metal. In future we might get the nodules contained much more of precious metals and raremetals.

The plants accumulate starch, lipid, and storage protein in their bud, seeds, and lumped stalk. The promoter of the gene coded storage protein is cloned and fused in frame to the gene coded heavy metal protein. The expression vector DNA is introduced into the plants and genetically modified organisms might accumulate heavy metals in storage tissue. It is easy to recover storage tissue. Starch in storage tissue is used as materials for manufacture and food, and processed to alcohol. Heavy metals with or in storage tissue are solubilized, recovered, and recycled. The elucidation of the symbiosis in plant–microbe and the construction of artificial symbiosis, the increasing efficiency of energy circulatory system will lead the development of environmental friendly metal-biotechnology.

References

1. Kruger, E. L., Anderson, T. A., and Coats, J. R., *Phytoremediation of Soil and Water Contaminants*, ACS symposium series, 664 (American Chemical Society, USA) (1997).
2. Hirsch, A. M., Lum, M. R., and Downie, J. A., *Plant Physiol.* **127,** 1484–1492 (2001).
3. Murooka, Y., *Nippon Nogeikagaku Kaishi* **77,** 146–149 (2003) (in Japanese).
4. Murooka, Y., Xu, Y., Sanada, H., Araki, M., Morinaga, T., and Yokota, A., *J. Ferment. Bioeng.* **79,** 38–44 (1993).
5. Hayashi, M., Maeda, Y., Hashimoto, Y., and Murooka, Y., *J. Biosci. Bioeng.* **89,** 550–553 (2000).
6. Yamashita, M., Kuwata, H., Murakami, H., and Murooka, Y., *J. Ferment. Bioeng.* **77,** 113–118 (1994).
7. Hong, S. H., Gohya, M., Ono, H., Murakami, H., yamashita, M., Hirayama N., and Murooka, Y., *Appl. Microbiol. Biotechnol.* **54,** 84–89 (2000).
8. Xu, Y. and Murooka, Y., *J. Ferment. Bioeng.* **80,** 276–279 (1995).

9. Sriprang, R., Hayashi, M., Yamashita, M., Ono, H., Saeki, K., and Murooka, Y., *J. Biotechnol.* **99,** 279–293 (2002).
10. Toyoma, M., Yamashita, M., Hirayama, N., and Murooka, Y., Interactions of arsenic with human metallothionein-2, *J. Biochem.* **132,** 217–221 (2002).
11. Sriprang, R., Hayashi, M., Yamashita, M., Ono, H., Saeki, K., and Murooka, Y., *Appl. Environ. Microbiol.* **69,** 1791–1796 (2003).
12. Ike, A., Sriprang, R., Ono, H., Murooka, Y., and Yamashita, M., *Chemosphere* **66,** 1670–1676 (2007).
13. Ike, A., Sriprang, R., Ono, H., Murooka, Y., and Yamashita, M., *J. Biosci. Bioeng.* **105,** 642–648 (2008).
14. Toyoma, M., Yamashita, M., and Murooka, Y., *J. Biosci. Bioeng.* **101,** 354 (2006).

Chapter 8

CONSTRUCTION OF YEAST BIOADSORBENT BY CELL SURFACE ENGINEERING

Kouichi Kuroda and Mitsuyoshi Ueda
Division of Applied Life Sciences, Graduate School of Agriculture, Kyoto University, Sakyo-ku, Kyoto 606-8502, Japan
k_kuro@kais.kyoto-u.ac.jp

8.1 Introduction

Pollution of soil and water along with the development of mining and manufacturing has emerged as a social issue. Particularly in the hydrosphere, heavy metal ion pollution has been caused by effluent wastewater from factories, mines, and metal refineries. This is a serious environmental issue to be addressed, but it is difficult to remove heavy metal ions from the environment because they cannot be chemically degraded, unlike organic pollutants. As a means of addressing this issue, there are two main methods of treating wastewater: physicochemical techniques and bioremediation using microorganisms as a bioadsorbent. Conventional methods of removing heavy metal ions by physicochemical techniques are

Handbook of Metal Biotechnology: Applications for Environmental Conservation and Sustainability
Edited by Michihiko Ike, Mitsuo Yamashita, and Satoshi Soda

www.panstanford.com

often ineffective and costly when applied to dilute effluents.[1,2] Recently, interest in bioremediation has increased owing to its great potential.[3–5] The cleanup of polluted wastewater using biological functions inherent in microorganisms is a prospective solution to the problem. In studies on bioremediation, how to recover and recycle heavy metal ions from a bioadsorbent after metal adsorption as well as removal of heavy metal ions are also problems to be solved. Therefore, a novel approach using a protein display system on the yeast cell surface (yeast cell surface engineering)[6,7] was applied to cell surface design for the molecular breeding of a bioadsorbent. In this chapter, yeast bioadsorbents constructed by displaying metal-binding molecules on their cell surface and their improvements are introduced, and the potential of an environmental cleanup system using the constructed yeast is discussed.

8.2 Metal Adsorption on Microbial Cell Surface

To recover metal ions present in aqueous water, adsorption of metal ions on an adsorbent in the first stage and effective desorption of adsorbed metal ions in the second stage are required. As an example of metal ion adsorption in aqueous water, adsorption of heavy metal ions was carried out to remedy environmental pollution. As a conventional material for metal adsorption, chemical resins such as polymers formed by a physicochemical method are extensively used. Recently, biotechnological adsorption of metals using microorganisms and plants has attracted much attention, because of its advantages in cost and potential of adsorption of low concentrations of metals.[3] Living organisms are equipped with a system for homeostasis to retain intracellular metal ions within a defined concentration, independent of the environmental concentration. In this system, there are functional proteins that recognize metal ions and mediate signal transduction, and sequester the toxic metal ions into nontoxic form. For the development of a bioadsorbent, the ability of biomolecules to recognize and bind metal ions is effectively utilized. Attempts to enhance the natural function of metal adsorption in living organism are actively carried out. Particularly for metal ions in aqueous solution, a bioadsorbent using microorganisms as an adsorbent material is expected as a prospective method. In a bioadsorbent, the adsorption phenomenon caused by the interaction between cells and metals is categorized into one of the following two processes. One is the adsorption on the cell surface, which is the first place to face and respond to the external environment, and the other is the adsorption by incorporation into the cell, namely, cellular accumulation. In the conventional development of a bioadsorbent, most research has been focused on the intracellular uptake, in which the cellular ability to uptake and accumulate metal ions has been improved.[8] However, the method of extracting the adsorbed metal ions from the cells has not been taken into consideration, and the system of cellular accumulation is insufficient in terms of the recovery of adsorbed metal ions. Cellular adsorption is limited to single use in this system because the disintegration of cells is required for the extraction of the metal ions accumulated inside cells. In contrast, the adsorption on the cell surface has significant advantages in that it proceeds in a short time and the adsorbed metal ions can be recovered without disintegration of cells. Therefore, the repeated use of cells for the adsorption and recovery of metal ions is possible. Microorganisms have a large surface area per unit volume and provide a large contact area for the interaction with metal ions in the surrounding environment. Furthermore, the adsorption ability of cells is maintained as long as the metal-binding proteins on the cell surface are functionally active, even if cells are nonliving. Metal adsorption on the microbial cell surface has the potential to be more suitable for biotechnological metal adsorption than intracellular adsorption because of the above-described advantages.

8.3 Biological Technique for Metal Adsorption by Novel Strategy

On the basis of the above-mentioned concept that adsorption is performed on the cell surface, the enhancement or endowment of metal adsorption ability on the cell surface would lead to the construction of a novel bioadsorbent. There has been no technology for this strategy up to now, and it is difficult to realize virtually. However, cell surface engineering, in which cell surface properties are designed by anchoring various functional proteins and peptides on the cell surface, is established and has attracted a great deal of attention in recent years for its wide range of applications.[6,7,9] Using this technology, cell surface adsorption has become realizable. Indeed, arming cells with new functions have already been constructed by cell surface engineering. In this technology, the display of proteins on the cell surface is achieved by fusion with cell surface proteins on the basis of molecular information. Many cell surface proteins have been identified for use in various microorganisms including *Saccharomyces cerevisiae*. *S. cerevisiae* has generally regarded as safe (GRAS) status, unlike other microorganisms. This is an important feature desirable for practical use. Among the yeast cell surface proteins, α-agglutinin is a mannoprotein involved in the sexual adhesion of mating-type **a** and α cells.[10] In the yeast display system, many heterologous proteins were successfully displayed using the N-terminal signal sequence and C-terminal cell-wall anchoring domain including the glycosylphosphatidylinositol anchor attachment signal of α-agglutinin (Fig. 8.1). Various heterologous proteins with comparatively large molecular masses have been successfully displayed with retention of their functions by the

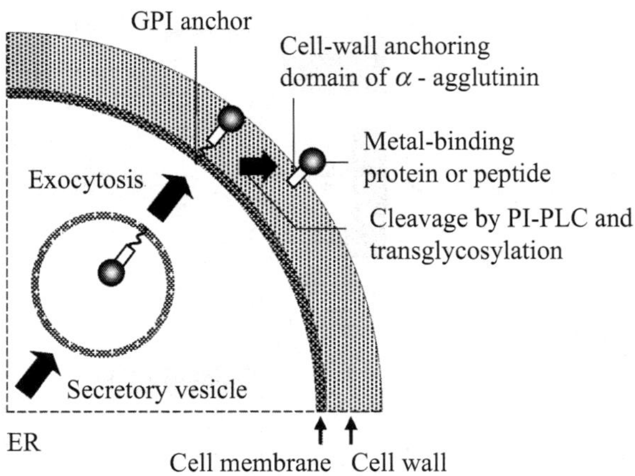

Figure 8.1. Mechanism of α-agglutinin-based molecular display on yeast cell surface.

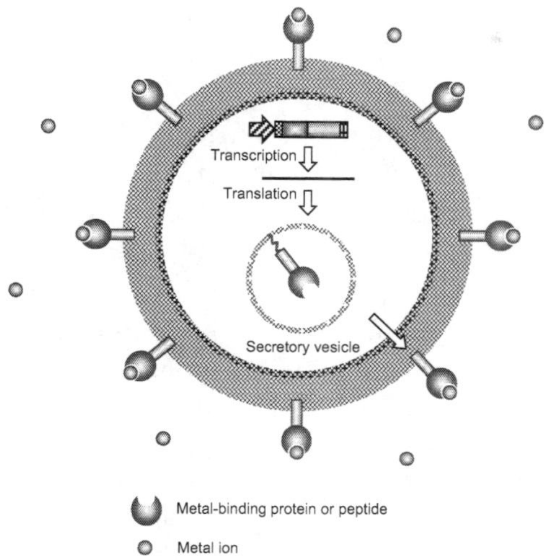

Figure 8.2. Model of yeast bioadsorbent constructed by cell surface engineering. See also Color Insert.

α-agglutinin-based display system in yeast.[6,7] Therefore, as a cell surface design for the adsorption and recovery of metal ions from aqueous solution, proteins and peptides with the ability to bind metal ions were displayed on the cell surface (Fig. 8.2).[11–14] In the construction of a bioadsorbent by cell surface engineering, a microorganism is used as a carrier of metal-binding molecules as well as a producer of molecules, and this leads to the simultaneous accomplishment of two processes, namely, the production and conjugation of a metal-binding molecule to a carrier. Furthermore, these processes can be automatically carried out by the convenient procedure of microbial culture. Thus, it is possible to prepare large amounts of a bioadsorbent in overnight cultivation, and to apply the cultivated cells directly to metal adsorption. Because the microorganism used in this case is a eukaryotic yeast, a quality control mechanism is equipped and all the metal-binding proteins and peptides with determined genome sequences could be displayed on the cell surface. Therefore, the construction of bioadsorbents for the adsorption and recovery of various metal ions by cell surface design is promising.

8.4 Cell Surface Display of Metal-Binding Proteins and Peptides on Yeast

Cell surface design for the construction of a bioadsorbent was performed by the display of proteins and peptides with the ability to bind metal ions. As the initial attempt, divalent heavy metal ions were targeted, and hexa-histidine peptide, [$(His)_6$; hexa-His], was used as a metal-binding peptide.[11] Hexa-His is widely used as an affinity tag in protein purification, and proteins fused with hexa-His can be purified by the affinity to divalent heavy metal ions such as nickel ions.[15] To display hexa-His on the yeast cell surface, the pre-pro leader sequence of the α-factor precursor,[16] hexa-His-encoding sequence, and 3′ half of the α-agglutinin gene were fused in flame from 5′ to 3′, and the fused sequence was regulated by a constitutive glyceraldehyde-3-phosphate dehydrogenase promoter[17] (Fig. 8.3). The constructed fusion gene was introduced and expressed in *S. cerevisiae*. The cell surface display of hexa-His was confirmed by immunofluorescence labeling of transformed cells. The observed fluorescence on the cell surface indicated that hexa-His was localized and displayed on the cell surface (Fig. 8.4). The constructed hexa-His-displaying yeast was applied to the adsorption of 100 μM heavy metal ions in aqueous solution. After the adsorption, cell surface-bound metal ions were recovered with ethylenediaminetetraacetic acid treatment. Hexa-His-displaying yeast showed the enhanced adsorption and recovery of copper and nickel ions, compared with control strains (Fig. 8.5). This result indicates that hexa-His displayed on the cell surface is functional and the cell surface design is effective in the strategy for the molecular breeding of a bioadsorbent. Furthermore, hexa-His-displaying yeast could grow

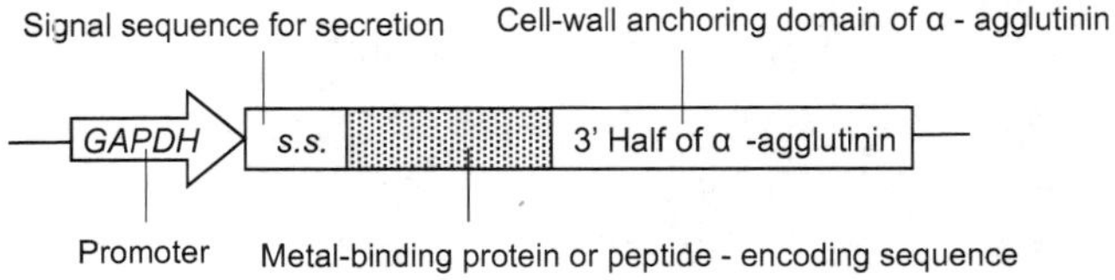

Figure 8.3. Fusion gene construction for cell surface display of metal-binding protein and peptide on yeast cell surface.

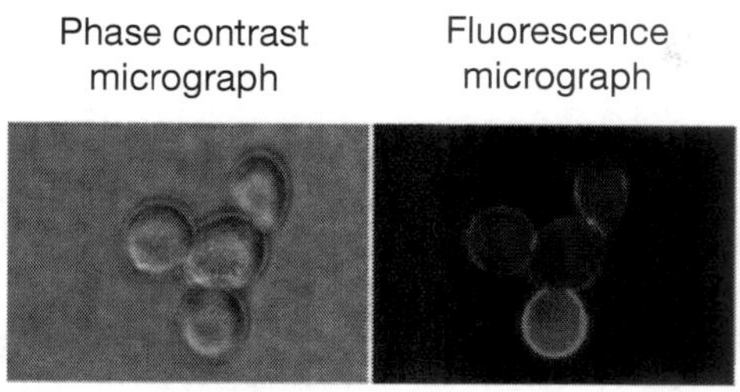

Figure 8.4. Confirmation of cell surface display of hexa-His by immunofluorescence labeling. See also Color Insert.

in the medium containing 2 mM copper ions, in which control strains could not grow. Therefore, the adsorption of copper ions by hexa-His on the cell surface led to enhanced tolerance to copper ions. This is an interesting result, which was not found in the case of hexa-His displayed on other microorganisms such as *Escherichia coli* and *Staphylococcus* species.[18,19]

Yeast metallothionein (YMT) was used as another metal-binding protein in the second attempt.[13] Metallothioneins are cystein-rich and ubiquitous proteins that bind and sequester heavy metal ions such as copper, cadmium, zinc, silver, and mercury. This protein plays an important role in the detoxification and storage of heavy

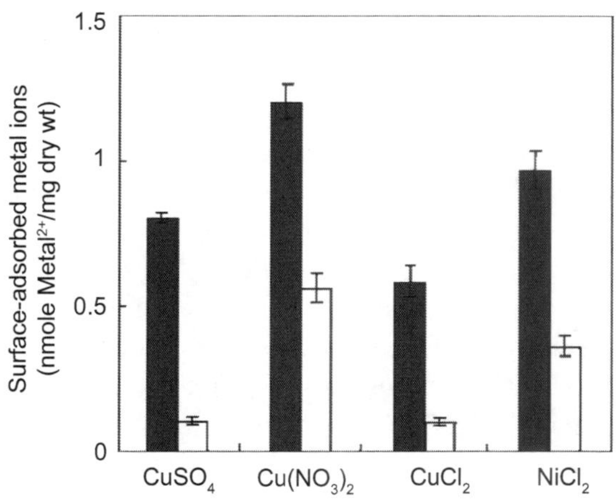

Figure 8.5. Adsorption and recovery of metal ions by hexa-His-displaying yeast. Gray bar: hexa-His-displaying yeast, white bar: control yeast.

metal ions for heavy metal homeostasis.[20,21] YMT was successfully displayed on the yeast cell surface by the α-agglutinin-based display system. From the result of cadmium ion adsorption using the constructed YMT-displaying yeast, the displayed YMT contributed to the improvement of cellular adsorption of cadmium ions. This strain also exhibited an enhanced tolerance to cadmium ions. Furthermore, YMT-displaying yeast could grow in the medium containing toxic cadmium ions, whereas hexa-His-displaying yeast showed tolerance to copper ions. The tolerance to metal ions enhanced by cell surface design is thought to be significant in that the bioadsorbent becomes applicable to a wider range of metal ion concentrations. Therefore, adsorption ability enhanced by cell surface engineering is effective in the molecular breeding of metal-tolerant cells as well as bioadsorbents.

8.5 Additional Improvement of Cell Surface-Engineered Yeast

In the bioadsorption of metal ions by cell surface-engineered yeast, yeast cells must be separated from treated water after the adsorption. Cells are commonly centrifuged for complete separation, but this requires equipment and is expensive. Therefore, the self-aggregation ability in response to the adsorption and accumulation of copper ions was appended to cell surface-engineered yeast as an additional improvement.[12] Cell aggregation is very important for the low cost and simplicity of the procedure because it causes cells to separate spontaneously from the treated water. Several genes have been demonstrated to be involved in the aggregation of yeast cells.[22] For addition of the aggregation ability, the fusion gene, which consists of a copper-inducible *CUP1* promoter from the YMT gene,[20] and *GTS1,* whose overexpression causes constitutive cell aggregation[23] (Fig. 8.6), was constructed and introduced into hexa-His-displaying yeast. The transformant containing the fusion gene aggregated in response to copper ions in culture medium. The aggregation response showed sensitivity to a copper ion concentration as low as 1 μM and rapid response within three hours.

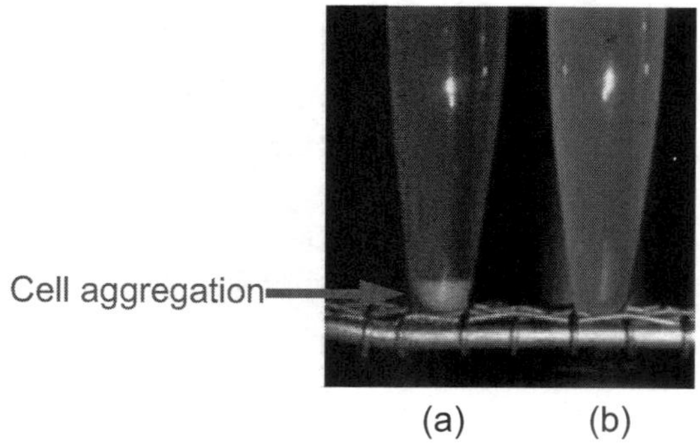

Figure 8.6. Self-aggregation of yeast cells harboring *CUP1* promoter-*GTS1* fusion gene in response to environmental copper ions. (a) Presence of 100 μM copper ions, (b) without copper ions. See also Color Insert.

These characteristics would be very useful in the separation of cell surface-engineered yeast after copper ion adsorption.

In the second improvement of cell surface-engineered yeast, YMT was tandemly fused and displayed by the α-agglutinin-based display system to enhance the adsorption ability.[14] Four and eight repeats as well as one YMT were successfully displayed on the yeast cell surface. The adsorption of cadmium ions depended on the number of displayed tandem repeats of YMT. Therefore, increasing the number of YMT repeats on the cell surface was an effective strategy for the improvement of the adsorption ability of a cell surface-engineered yeast, and the characteristics of cell surface-engineered yeasts as a bioadsorbent were dependent on the ability of the displayed metal-binding proteins.

8.6 Conclusion

Cell surface design directed to metal ion adsorption on cell surface was attempted for the novel strategy of bioadsorption. The constructed surface-engineered yeasts showed superior features compared with conventional bioadsorbents. In the adsorption system of metal ions using the cell surface, it is possible to desorb and recover the metal ions after bioadsorption by mild chemical manipulation. This allows the recycling of adsorbed metal ions and the reutilization of bioadsorbents leading to economical advantages, since there is no need for cell disruption in the recovery of adsorbed metal ions. Furthermore, metal ion adsorption on the cell surface provided cellular tolerance to toxic metal ions. These are valuable characteristics for practical use. By introducing an artificial signal transduction pathway into surface-engineered yeast, the easy separation of cells after metal adsorption by self-aggregation in response to metal ions was achieved. In addition, the adsorption ability of surface-engineered yeast was enhanced by tandem repeating of a metal-binding protein displayed on the cell surface. Therefore, the adsorption, recovery, and recycling system of metal ions using cell surface-engineered yeast constructed by cell surface engineering is promising for next-generation bioadsorption.

References

1. Marques, A. M., Roca, X., Simon-Pujol, M. D., Fuste, M. C., and Congregado, F., *Appl. Microbiol. Biotechnol.* **35**, 406–410 (1991).
2. Kapoor, A. and Viraraghavan, T., *Bioresour. Technol.* **53**, 195–206 (1995).
3. Gadd, G. M. and White, C., *Trends Biotechnol.* **11**, 353–359 (1993).
4. Lovely, D. R. and Coates, J. D., *Curr. Opin. Biotechnol.* **8**, 285–289 (1997).
5. Eccles, H., *Trends Biotechnol.* **17**, 462–465 (1999).
6. Ueda, M. and Tanaka, A., *Biotechnol. Adv.* **18**, 121–140 (2000).
7. Kondo, A. and Ueda, M., *Appl. Microbiol. Biotechnol.* **64**, 28–40 (2004).
8. Pazirandeh, M., Chrisey, L. A., Mauro, J. M., Campbell, J. R., and Gaber, B. P., *Appl. Microbiol. Biotechnol.* **43**, 1112–1117 (1995).
9. Georgiou, G., Poetschke, H. L., Stathopoulos, C., and Francisco, J. A., *Trends Biotechnol.* **11**, 6–10 (1993).
10. Lipke, P. N. and Kurjan, J., *Microbiol. Rev.* **56**, 180–194 (1992).
11. Kuroda, K., Shibasaki, S., Ueda, M., and Tanaka, A., *Appl. Microbiol. Biotechnol.* **57**, 697–701 (2001).
12. Kuroda, K., Ueda, M., Shibasaki, S., and Tanaka, A., *Appl. Microbiol. Biotechnol.* **59**, 259–264 (2002).
13. Kuroda, K. and Ueda, M., *Appl. Microbiol. Biotechnol.* **63**, 182–186 (2003).

14. Kuroda, K. and Ueda, M., *Appl. Microbiol. Biotechnol.* **70**, 458–463 (2006).

15. Houchi, H., Dobeli, H, and Schacher, A., *J. Chromatogr.* **411**, 177–184 (1987).

16. Takahashi, S., Ueda, M., Atomi, H., Beer, H. D., Bornscheuer, U. T., Schmid, R. D., and Tanaka, A., *J. Ferment. Bioeng.* **86**, 164–168 (1998).

17. Sawai-Hatanaka, H., Ashikari, T., Tanaka, Y., Asada, Y., Nakayama, T., Minakata, H., Kunishima, N., Fukuyama, K., Yamada, H., Shibano, Y., and Amachi, T., *Biosci. Biotechnol. Biochem.* **59**, 1221–1228 (1995).

18. Sousa, C., Cebolla, A., and de Lorenzo, V., *Nat. Biotechnol.* **14**, 1017–1020 (1996).

19. Samuelson, P., Wernerus, H., Svedberg, M., and Stahl, S., *Appl. Environ. Microbiol.* **66**, 1243–1248 (2000).

20. Butt, T. R. and Ecker, D. J., *Microbiol. Rev.* **51**, 351–364 (1987).

21. Perego, P. and Howell, S. B., *Toxicol. Appl. Pharmacol.* **147**, 312–318 (1997).

22. Teunissen, A. W. and Steensma, H. Y., *Yeast* **11**, 1001–1013 (1995).

23. Bossier, P., Goethals, P., and Rodrigues-Pousada, C., *Yeast* **13**, 717–725 (1997).

Chapter 9

BIOLEACHING OF METALS FROM SOLID WASTE INCINERATION ASH

Tomonori Ishigaki
Center for Material Cycles and Waste Management Research, National Institute for Environmental Studies, 16-2 Onogawa, Tsukuba, Ibaraki, 305-8506, Japan
ishigaki@nies.go.jp

9.1 Waste Incinerataion Ash

Waste incineration has been widely adapted as an effective means to sanitize and to reduce the volume of waste before landfilling in several countries such as Japan, Switzerland, and Sweden. Air pollution, including dioxin emission, and some issues on the waste incineration had been already cleared by the effort of technology improvement under the strict regulations in these countries.[1] On the other hand, the recent movement for sustainable waste management has also promoted the effective utilization of waste incineration ash. The secondary usage of waste incineration ash from municipal solid waste incineration (MSWI) plants has been suggested as an "artificial ore"[2] or for use as a construction material

Handbook of Metal Biotechnology: Applications for Environmental Conservation and Sustainability
Edited by Michihiko Ike, Mitsuo Yamashita, and Satoshi Soda

www.panstanford.com

by cementation or vitrification,[3] but a large portion of the ash is still disposed of in waste landfill sites.

Fly ash, which is included in combustion gas and is recovered at the dust collection process, is an anthropogenic concentrate containing a wide variety of toxic and valuable heavy metals of low boiling point such as lead, zinc, and cadmium. These concentrations are comparable to content in ore that has been subjected to mining.[4] It has become very important to recover and utilize these valuable metals, because most of these metal resources are strictly competitive for resource securement under the global industrialization.[1,4,5] However, current main stream of fly ash treatment is a cementation, a chemical immobilizing treatment, and a melting before landfilling under the requirement of National or regional regulations. Although these processes are adapted from the viewpoint of the environmental safety thorough the containment of toxic metals, it was also reported that the long-term investigation of metals in a waste landfill showed the possibility of re-mobilization by the change of environmental condition in waste landfill.[6,7]

9.2 Recovery of Metals from Solids

Technology used for recovering metals from solids was generally diverted from the mining technology of physico-chemical methods such as thermal desorption, chemical leaching, and liquid elution, or biological method.[8,9] Biological method had gained the much attention because it requires less energy and could be applicable for the materials containing low concentration of metals.[10,11] Several biological technologies of metal recovery were indicated such as biosorption, biomineralization, biovolatilization, and bioleaching. Amongst of these technologies, the bioleaching is only able to apply for recovering the various metal elements from solid phase, and others would object the recovery of limited metal from liquid phase. Although the bioleaching is one of the most promising technologies, few practice of bioleaching from the solids except ore has been reported. It can be considered that the bioleaching from MSWI ash must be under the technology development for the practical use.

9.3 Bioleaching

9.3.1 *Overview of Bioleaching*

The bioleaching must be defined to solubilize the metal from the solid phase by means of biological process. In this sense, the bioleaching can be applied for the remediation of the metal-polluted soil. However, from the viewpoint of the cost-benefit performance, the bioleaching has been normally recognized as the technology for recovering the valuable resources.[12] Practically this technology was applied in the mining: the addition of the bacterial culture and media to the unmineable ore (*in situ* bioleaching) or mined residues of ore (dump leaching, heap leaching). Secondary recovery of metals by the bioleaching would be only feasible for the ore containing the metals of rarity value, such as gold, silver, and arsenic.[13,14] Since the bioleaching in the incubation reactor could cost rather expensive, it must be applied for the ore containing low content but high marketable metals such as copper and uranium.[15]

Bioleaching using a fungal microorganism producing gluconic acid or citric acid, sulfur-oxidizing bacteria producing sulfuric acid, iron-oxidizing bacteria oxidizing a reduced metal compounds, and iron-reducing bacteria reducing ferric iron to ferrous iron has been reported as a promising technology for recovering valuable metals from fly ash,[16] because the bioleaching is cleaner and consumes less energy than pyro- or hydrometallurgical methods. Sulfur- and iron-oxidizing bacteria are the autotrophic bacteria; sulfur-oxidizing bacteria such as *Thiobaclillus thiooxidans* or *Acidithiobacillus caldus* can utilize elemental sulfur or sulfur oxides (S^{2-}, $S_2O_3^{2-}$, SO_3^{2-}) and iron-oxidizing bacteria such as *Thiobaclillus ferooxidans* or *Leptospirillum ferooxidans* can utilize both sulfuric compounds and the ferrous iron as their energy source under the oxidative condition. Since the optimal pH range for growth of sulfur- and iron-oxidizing bacteria is 1–5, the strict oxidative condition produced by them inhibits the growth of bacteria other than sulfur- and iron-oxidizing bacteria. Further, additional operational cost for the cultivation might be cut down because carbon dioxide will be used as their carbon source. Although the bioleaching by heterotrophic bacteria was also reported, the requisite of the organic carbon source and/or the competition in microbial community were remained as the problems for achieving the effective leaching.[2,17]

9.3.2 *Mechanisms of Bioleaching*

In the bioleaching process, it has been generally considered that metals could be mobilized by mechanisms of a direct enzymatic action and an indirect action mediated by the bacterial activity such as redox potential or inorganic acid formation.

Direct enzymatic action is considered to be the cabalistic reaction by the bacteria attached onto ore surface as following reaction,

$$MeS + 2O_2 + 2H^+ \rightarrow Me^{2+} + H_2SO_4, \tag{9.1}$$

where Me is a bivalent metal.[16]

This reaction depends on the electron transfer thorough the formation of a fine battery between the bacterial cell produced by enzyme and amorphous sulfide ore. The reaction shown in Eq. 9.1 will be occurred for CuS, CuS_2, ZnS, PbS, MoS_2, CoS, NiS, Sb_2S, and Sb_2S_3.[18,19]

Further, Sand *et al.*[20] proposed the plausible mechanism of metal leaching that essentially proceeded from ferric iron complexed by secreted extracellular polymeric substances. This concept of a "contact" leaching mechanism has been generally accepted and become popular, and it appeared that the direct leaching mechanism might not exist.

On the other hand, the indirect bioleaching action is the ionization of metals by oxidizing agent such as sulfate, ferric iron, or organic acid which were generated by the biological reaction. Representative reactions will be expressed by following equations:

$$4Fe^{2+} + O_2 + 4H^+ \rightarrow 4Fe^{3+} + 2H_2O \tag{9.2}$$

$$MeS + 2Fe^{3+} \rightarrow Me^{2+} + 2Fe^{2+} + S^0 \tag{9.3}$$

$$2S^0 + H_2O + 3O_2 \rightarrow 2H_2SO_4 \tag{9.4}$$

$$MeS + H_2SO_4 \rightarrow Me^{2+} + 2SO_4^{2-} + S^{2-} \tag{9.5}$$

Equations 9.2 and 9.3 show the oxidation of ferrous iron and the metal solubilization by ferric iron produced by iron-oxidizing bacteria, respectively. Equations 9.4 and 9.5 show the oxidation of elemental sulfur and the metal solubilization by sulfate produced by both sulfur- and iron-oxidizing bacteria, respectively. Metal is known to be stable as ionized state under the high redox potential and low pH. The indirect bioleaching is the solubilization of metal by this condition that was produced by bacteria. Bioleaching of metal sulfide by iron-oxidizing bacteria (Eqs. 9.2 and 9.3) also generates elemental sulfur. It is practical problem that the elemental sulfur attached on the surface of sulfide ore must inhibit the effective successive bioleaching. However, co-culture of sulfur-oxidizing bacteria and iron-oxidizing bacteria support to continue the bioleaching by the solubilization of this elemental sulfur by the reaction shown in Eq. 9.4.

9.3.3 *Bioleaching of Metals from MSWI ash*

As mentioned above, the bioleaching of solids has different two mechanisms of direct and indirect actions. However, it was widely known that the direct mechanism is only exhibited for the sulfide state of metal. On the other hand, most of metals in MSWI ash had known to be nonsulfide such as oxide or carbonate state. Previous studies[22,23] reported that the nonsulfide ore included uranium, copper, and manganese would be solubilized by the indirect bioleaching mechanisms. In contrast, metals in the waste landfill should be often transformed to sulfide state under the reducing condition of waste layer. These metal sulfides would be expected to be solubilized by the bioleaching as well as sulfide ore.

Even these reports suggested that for the efficient bioleaching, less attention has been paid to the possible effectiveness of bioleaching of solid wastes, including fly and bottom ash from MSWI.

Some previous reports indicated that the bioleaching of MSWI fly ash required a lot of time, resulting in limited leachability.[17,24,25] Recognizing the difficulty, further research and development to improve the efficiency of metal recovery is needed before we would apply the bioleaching to full-scale waste landfills.

Tateda *et al.*[24] reported the sulfur-oxidizing bacterium *T. thiooxidans* incubated with 0.7% of MSWI ash resulted in the solubilization of cadmium, copper, and zinc, more than 50% of its content. However, the addition of MSWI ash with 3% of culture media significantly decreased the leachability of these elements. Increase of MSWI ash must inhibit the formation of the acidic condition because of high alkali compounds included in MSWI ash. The crucial point for the practical development of effective bioleaching will be the formation and maintenance of the optimum condition for the bioleaching. Krebs *et al.*[17] could achieve the high leachability such as more than 80% of cadmium, copper, and zinc, 60% of aluminum, and more than 30% of iron and nickel using the culture of *T. thiooxidans* with addition of 8% of MSWI ash and the semi-continuous inoculation. Further, the utilization of *T. ferooxidans* as iron-oxidizing bacterium and the preliminary pH adjustment had shown the effective leachability of cadmium (90%), zinc (80%), and lead (70%) at the addition of solid waste mixture of 10% to the culture medium.[25]

Effective leaching from ore using a mixed culture of sulfur-oxidizing bacteria and iron-oxidizing bacteria has been reported intensively.[26,27] It has been believed that the application of this technology to fly ash will improve its leachability. However, earliest trials of bioleaching from fly ash have focused on the utilization of a pure culture of sulfur-oxidizing bacteria or iron-oxidizing bacteria, with only one published report of an attempt to bioleach the fly ash using a mixed culture.[16]

Brombacher *et al.* examined that the semi-continuous bioleaching plant consisted of three serially connected reactors, the MSWI ash suspension vessel, and the bacterial stock culture incubator.[16] Ash addition ratio was 10% of the reactors, and leaching efficiencies for cadmium, copper, zinc, nickel, and aluminum were 100%, 90%, 80%, 60%, and 50%, respectively.

Further, Ishigaki *et al.*[28] reported that the simultaneous reaction of the iron oxidation in the mixed culture of sulfur-oxidizing bacteria and iron-oxidizing bacteria must lead the condition of high redox potential and low pH. It could effectively enhance the bioleachng by the inhibition of the increase of pH in the culture by addition of MSWI ash. Amongst of the leached metals, chromium and arsenic would be solubilized by the ferric iron mechanism, and copper, zinc, and cadmium would be solubilized by the sulfate mechanism.

Detailed investigation of the mechanism of metal mobilization from ash by a mixed culture is therefore requisite. Further, to control successfully the bioleaching process as the *in situ* technology at the waste landfill or the pretreatment technology before landfilling, we must investigate the control factors of the engineering process operation.

9.4 Bioleaching of/from MSWI Ash in Waste Landfills

To consider the feasibility of the bioleaching of metals from MSWI ash before landfilling, not only the function of the removal of hazardous metals but the recovery of valuable metal component must be included in the total system. However, current situation on the development of sound-material cycle society will orient the up-stream separation or segregation and decrease of the amount of the metal input to the incinerators. It can be considered that the some applicability of the bioleaching technology should be small but be remained in the certain field. For example, excavated waste from landfills under the remediation project must be objected to remove and to recover the metals. Further, *in situ* bioleaching for waste landfills could be expected as an analogue of mining technology. It must be expected to apply for the field that could show high cost-benefit efficiency, e.g. highly hazardous and residents health.

Few examination of the *in situ* bioleaching of metals from the landfill had been reported even in the experimental scale. As mentioned above, the *in situ* bioleaching must be based on the technology using the mining, and the method of the supplement of nutrients and electron acceptors (oxygen for sulfur- and iron-oxidizing bacteria) must be also investigated with referring the bioremediation for the polluted soil and/or groundwater.

The study on metal removal from MSWI ash using a lab-scale landfill reactor that simulated the ash monofill type landfill have examined the effect of irrigation of the preculture of mixed culture of sulfur- and iron-oxidizing bacteria with ferrous iron and elemental sulfur as nutrients.[29] It had proved that the effective bioleaching had occurred with the application of the air sparging at the saturated zone. This effective bioleaching was also supported by the decrease of pH in waste layer due to the leachate circulation. Leaching activity had continued for 90 days, and this long-term bioleaching achieved the high leachability for chromium (54%), copper (39%), cadmium (29%), and manganese (22%). Further, this report also revealed the influence of environmental characteristics of waste landfills. Presence of easily degradable organics resulted to inhibit the activity of iron-oxidizing bacteria by the supporting effect on heterotrophic bacteria. And cover soil layer would exhibit the buffer effect on the pH increase caused by MSWI ash. These factors of inhibition/stimulation for bioleaching must be the information on practical application for waste landfills.

9.5 Conclusion

Although there are several keywords on future society such as "sustainable development" or "sound-material cycle society," waste landfilling will be one of the ultimate options without alternatives from the viewpoint of the achievement of the public health and the appropriate management of waste. Against the situation on the global transportation of the resource and products, the movement of resource nationalism must generate the next hot issue on the material cycles including the waste management. Each country/region should recognize the potential of MSWI ash in which metal was accumulated thorough the anthropogenic activity. Ash itself and the waste landfills containing ash must be re-evaluated from the viewpoint of not only environmental safety but also resource productivity as an artificial ore and mine. Several technical methodologies such as the bioleaching described in this chapter will be prepared for the reasonable and feasible potential solution of the coming metal crisis near future.

References

1. Tavares, G., Zsigraiova, Z., and Semiao, V., *Proc. Sardinia 2009, Twelfth International Waste Management and Landfill Symp* (2009).
2. Bosshard, P. P., Bachofen, R., and Brandl, H., *Environ. Sci. Technol.* **30**, 3066–3070 (1996).
3. Ferreira, C., Ribeiro, A., and Ottosen, L., *J. Hazard. Mater.* **96**, 201–216 (2003).

4. Kokado, M., *Metallurgical Rev. MMIJ* **16**, 105–119 (1999).
5. Hara, Y. and Kitahashi, S., *Proc. 8th Symp. Geo-environ. Geo-technics* 221–224 (1998).
6. Bozkurt, S., Moreno, L., and Neretnieks, I., **250**, 101–121 (2000).
7. Jung, K., Chang, Y. C., Takamizawa, K., and Yoo, Y. S., *Jpn. J. Water Treat. Biol.* **35**, 143–155 (1999).
8. Mulligan, C. N., Yong, R. N., and Gibbs, B. F., *Eng. Geol.* **60**, 193–207 (2001).
9. Soil Environmental Rehabilitation Association in Japan, *Current Situation and Countermeasure on Soil and Groundwater Pollution* (Ohmsya, Japan) (2000).
10. John, T. C., *Bioremediation Engineering, Design and Operation* (McGraw-Hill, USA) (1995).
11. Fujita, M., Yagi, O., and Urushigawa, Y., *Handbook of Practical Bioremediation* (Realize Science & Engineering Center, Japan) (2001).
12. Means, J. L. and Hinchee, R. E., *Emerging Technology for Bioremediation of Metals* (Lewis Publishiers, USA) (1994).
13. Attia, Y. A. and Elzeky, M., *Miner. Metall. Process* **8**, 122–127 (1991).
14. Wan, R. Y. and Brierley, J. A., *Miner. Eng.* **49**, 76–80 (1997).
15. Ehrlich, H. L., *Miner. Metall. Process* **5**, 57–60 (1988).
16. Brombacher, C., Bachofen, R., and Brandl, H., *Appl. Environ. Microbiol.* **64**, 1237–1241 (1998).
17. Krebs, W., Bachofen, R., and Brandl, H., *Hydrometallurgy* **59**, 283–290 (2001).
18. Imai, K., *Cell Sci.* **2**, 496–502 (1986).
19. Silver, M. and Torma, A. E., *Can. J. Microbiol.* **20**, 141–147 (1974).
20. Sand, W., Gehrke, T., Jozsa, P. G., and Schippers, A., *Hydrometallurgy* **59**, 159–173 (2001).
21. Nagamori, M. and Ono, Y., *Annual Report of Saitama Institute of Environmental Pollution* **21**, 49–65 (1994).
22. Zajic, J. E., *Microbial Biogeochemistry* (Academic Press, UK) (1969).
23. Chida, T., *Microbiology and Its Applications to Resource Engineering* (Corona Publishing Co., Ltd., Japan) (1996).
24. Tateda, M., Ike, M., and Fujita, M., *Resource Environ. Biotechnol.* **2**, 137–151 (1998).
25. Mercier, G., Chartier, M., Couillard, D., and Blais, J. F., *Environ. Manag.* **24**, 517–528 (1999).

26. Curutchet, G., Tedesco, P., and Donati, E., *Biotechnol. Lett.* **18**, 1471–1476 (1996).
27. Giaveno, A. and Donati, E., *Process Biochem.* **36**, 955–962 (2001).
28. Ishigaki, T., Nakanishi, A., Tateda, M., Ike, M., and Fujita, M., *Chemosphere* **60**, 1087–1094 (2005).
29. Ishigaki, T., Nakanishi, A., Tateda, M., Ike, M., and Fujita, M., *Proc. 12th annual conference of the Japan Society of Waste Management Experts* **I**, 382–384 (2001) (in Japanese).

Chapter 10

DISSOLUTION OF PRECIOUS METALS FROM WASTE PRINTED CIRCUIT BOARDS BY USING BACTERIA

Tadashi Takemoto
Joining and Welding Research Institute, Osaka University, ZIP 567-0047, 11-1 Mihogaoka, Ibaraki-shi, Osaka Japan
takemoto@jwri.osaka-u.ac.jp

10.1 Introduction

The development of low environmental impact industrial technology is highly expected to reduce the rate of global warming, because various manufacturing and process engineering using petroleum resources has been believed to enhance environmental impact by exhausting CO_2 and other global warming substances. Utilization of microorganism such as bacteria to industrial processing is one of the solutions to promote the environmentally friendly engineering.[1] Many bacteria had been utilized in fermentation industries from old times and recently it has been developing to decompose the toxic substances and purging of polluted area.[2] The development of new application of microorganism to industrial engineering is growing as environmentally friendly low impact process in recent days using many functions of bacteria. The major interest of bioleaching using

Handbook of Metal Biotechnology: Applications for Environmental Conservation and Sustainability
Edited by Michihiko Ike, Mitsuo Yamashita, and Satoshi Soda

www.panstanford.com

bacteria is leaching of several metallic elements such as Cu,[3–8] Ni,[9] and several heavy metals such as Cd, Pb, and Zn[10–12] from minerals and waste substances.

Staphylococcus causes microbial corrosion of several metals and alloys.[13–15] Microbial corrosion is not a good phenomenon in industrial view point; however, the interaction between bacteria and metals can be positively used in metallic refining and leaching processes. In fact, the leaching technology of toxic metals from several residues by using bacteria is also proposed as environmental friendly technology.[11]

The reaction rate between bacteria and metal is extremely low; however, the study to enhance the reaction was also conducted in bioleaching of Cu.[16,17] From the viewpoint of resource strategy, it is useful to establish the recycle technology of precious metals from waste electronics such as printed wiring boards (PWBs) mounting various electronic components containing several precious metals such as Au, Ag, Pd, etc.[18] Especially, Au plating thickness is reducing year by year according to the development of thin film technology; therefore, the recycle amount of gold from waste printed circuit boards (PCBs) is reducing. It means that the recycling of Au from PWBs doesn't fit for the current industrial process from the economical viewpoint. However, to recycle precious elements from PCBs is useful because these metals are exhaustive elements. The following describes the ability of cyan generating bacteria to recover valuable metals from PWBs and the method to enhance the reaction rate between bacteria and metals.

10.2 Dissolution of Au by Cyan Generating Bacteria

Table 10.1 shows the cyan generating ability of four bacteria in YP culture medium indicated in the table. Among four bacteria, two of them showed no cyan. *Chromobacterium violaceum* showed the best result generating 1.0 mmol/L of cyan. In this culture medium the maximum concentration of 1.0 mmol/L was obtained after 50 h of cultivation, and the concentration gradually reduced to 0.3 mmol/L after 400 h of cultivation. *Burkholderia cepcia* also produced cyan; however, the maximum concentration was 0.2 mmol/L. Accordingly,

Table 10.1. Cyan generating bacteria and composition of YP culture used for dissolution test of Au.

		Maximum generated cyan concentration (mmol/L)
Culture condition	Incubation under forced circulation (30°C, stirring rate: 110 rpm)	
Composition of culture	Polypepton: 10 g, yeast extract: 2 g, $MgSO_4 \cdot 7H_2O$: 1 g, distilled water: 1 L	
Bacteria	*Burkholderia cepcia*	0.2
	Chromobacterium violaceum	1.0
	Pseudomonas chlororaphis	0.0
	Pseudomonas fluorescens	0.0

the most appropriate bacterium was found to be *C. violaceum* in culture medium indicated in Table 10.1.

Figure 10.1 shows the dissolved concentration of Au powder in YP culture using four bacteria indicated in Table 10.1. The dissolution of Au was only observed in medium containing cyan generated bacteria. *C. violaceum* producing the maximum cyan concentration showed the highest dissolved Au concentration, 0.4 mmol/L of Au after 400 h, where the used Au powder content corresponded to 1 mmol/L. *B. cepcia* also dissolved Au powder; however, the dissolved Au concentration was less than 0.1 mmol/L after 400 h incubation. Of course no dissolution of Au was observed in bacteria producing no cyan. It is clear that Au can be dissolved in media containing cyan that was produced by bacteria.

Figure 10.2 shows the change of cyanide concentration together with cell population of *C. violaceum* in YP culture. The change of cell population shows the normal process, growth phase, stationary phase, and death phase. Cyan concentration showed sudden increase at early stage of incubation and then rapidly decreased, and finally the concentration became a constant in death phase. The initial increase of cyan well corresponded with the increase of cell population, indicating that *C. violaceum* generated cyan actively at early stage of incubation.

Cyan is formed by the following reaction.

$$NH_2CH_2COOH \rightarrow HCN + CO_2 + 4[H] \qquad (10.1)$$

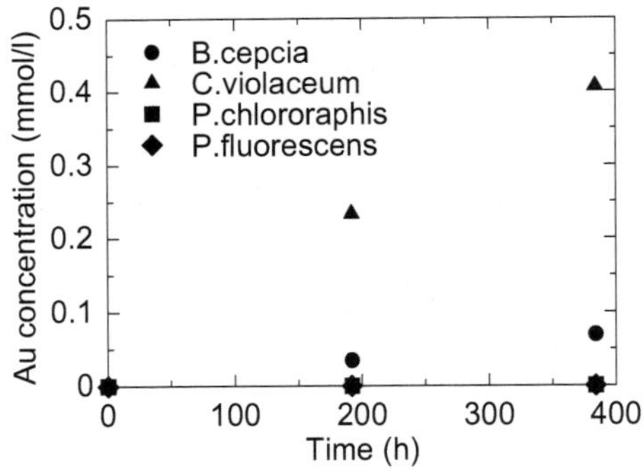

Figure 10.1. Relation between cultivation time and dissolved Au concentration in culture medium with different bacteria.

The decrease of cyan concentration after reaching the maximum cell population is based on the following decomposition reaction.

$$NH_2CH(COOH)(CH_2)_2COOH + HCN \rightarrow NH_2CH(COOH)(CH_2)CH_2CN + CH_3COOH \qquad (10.2)$$

$$NH_2CH(COOH)(CH_2)CH_2CN + H_2O \rightarrow NH_2CH(COOH)CH_2CONH_2 \qquad (10.3)$$

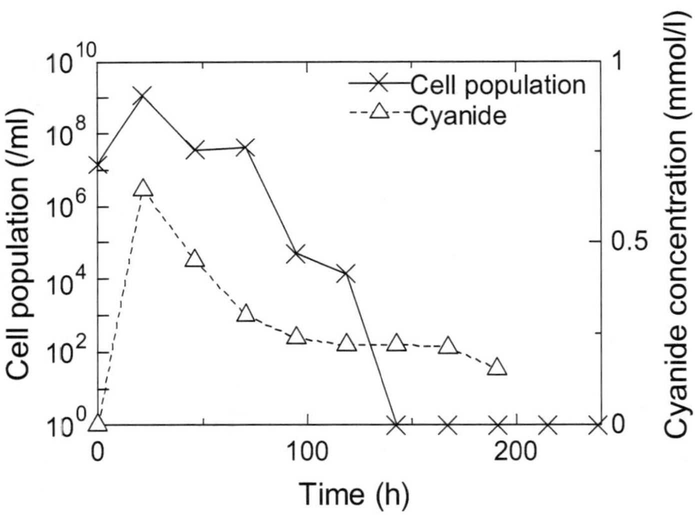

Figure 10.2. Changes of cell population and cyanide concentration in culture medium with *C. violaceum*.

The decrease of cyan is also based on the decrease of polypeptone in culture medium due to the reaction (1) and the self-defense ability of bacteria to guard itself from the toxic substances. This decomposition reaction becomes a key for the environmentally conscious processing because the reaction decomposes the toxic cyan.

10.3 Dissolution of Au from Waste PWBs

The results of dissolution test on waste printed wiring boards (PWBs) are shown in Fig. 10.3. The constitution of PWBs is indicated in Table 10.2; it consisted of Au/Ni/Cu from the surface. The surface of PWB is clearly damaged after 288 h test in bacteria added culture medium and Au plating was almost completely disappeared after 480 h test. On the other hand, in culture solution without bacteria for comparison, the dissolution of surface Au was not observed. To clarify the dissolution of Au simply, quantitative energy dispersive X-ray spectroscopy (EDX) was conducted on tested PWBs.

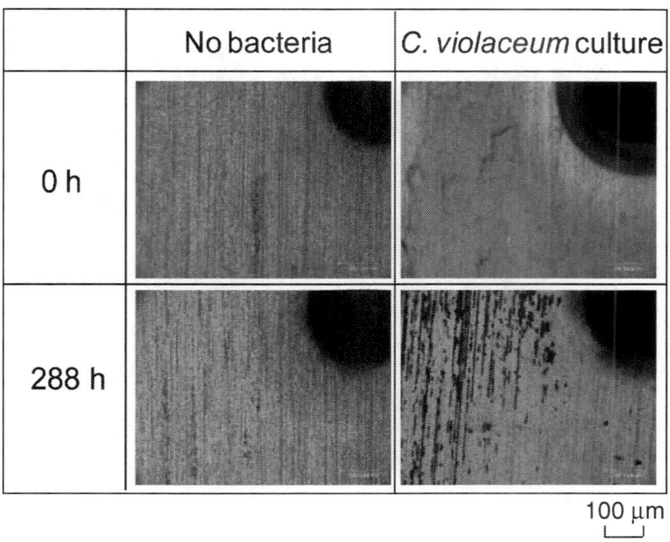

Figure 10.3. Change of surface morphology of Au plated Cu conductor on PWB immersed in culture medium with and without *C. violaceum*. See also Color Insert.

Table 10.2. Constitution of printed wiring board.

	Materials and constitution
Base material	Glass-epoxy resin: FR-4
Conductor	Cu: 35 μm in thickness
Plating	Au/Ni: Au surface finish of 0.07 μm thickness, Ni of 4 μm thickness

Figure 10.4 shows the result of quantitative analysis on several test conditions. Of course the results only show the qualitative tendency of dissolution of plating because the thickness of plating is limited; however, it is clear that Au concentration decreased with test period. The concentration of Cu conductor increased with test time that indicated the dissolution Ni plating. After the test period of 480 h it is clear that Ni plating was almost completely dissolved. On the other hand, in the culture solution without bacteria, Au concentration was kept constant value that means the no dissolution of Au in this culture solution.

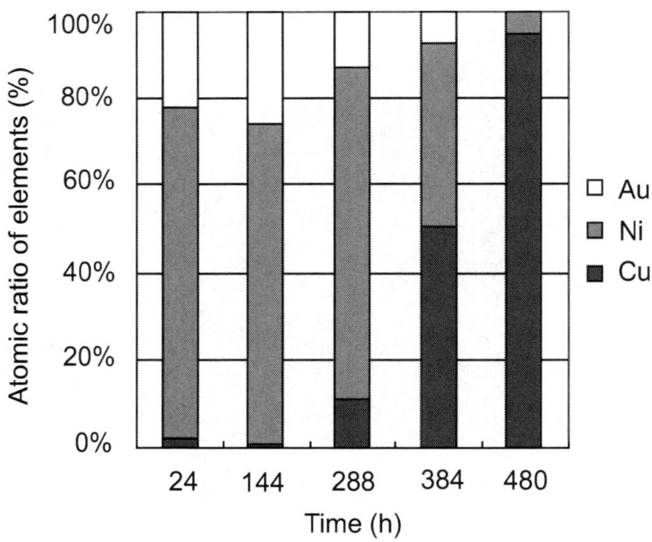

Figure 10.4. Effect of immersion time in culture medium with *C. violaceum* on results of EDX analysis of surface of PWB conductor.

10.4 Increase of Dissolution Rate of Au

The dissolution reaction of Au is controlled by electrochemical reaction; therefore, to enhance the dissolution rate of Au activation of both the anode reaction of Au dissolution and the cathode reaction consuming oxygen is necessary. It is desirable to increase both cyan and oxygen concentration in culture medium; however, oxygen is drastically consumed by active increase of cell population, because the bacteria consumes oxygen to survive. Figure 10.5 shows the change of oxygen concentration in culture during test time. The oxygen concentration drastically decreased during the active growth phase. The Au dissolution reaction proceeds under the oxygen depleted condition; the state is cathode reaction rate controlled process. Accordingly the supply of oxygen to culture is expected to enhance the dissolution rate of Au effectively.

Figure 10.6 shows the effect of dissolved oxygen concentration on dissolution of Au.[19]

Both aeration and H_2O_2 addition gave little influence on generated cyan concentration. Addition of H_2O_2 increased dissolved oxygen concentration even at growth phase. Aeration gave no increase of dissolved oxygen at growth phase; however, dissolved oxygen concentration was increased at stationary phase, after 20 h. Under these conditions, dissolved Au concentration was high in H_2O_2 added culture having high oxygen. Dissolved Au increased at stationary phase, after 20 h, where dissolved oxygen was increased under aeration condition.

Figure 10.7 shows the relation between dissolved oxygen concentration and Au dissolution rate.[19] The rate almost linearly increased with increase of dissolved oxygen irrespective of the test conditions. The relation clearly indicates that dissolved oxygen concentration controls the Au dissolution rate independent of experimental condition. This fact means that Au dissolution rate in culture was controlled by cathode reaction rate due to the depletion of dissolved oxygen in culture. Accordingly the supply of oxygen in culture is quite important to enhance Au dissolution rate; however, aeration at growth phase has no positive effect. On the other hand, the addition of H_2O_2 is effective to increase dissolved oxygen and to

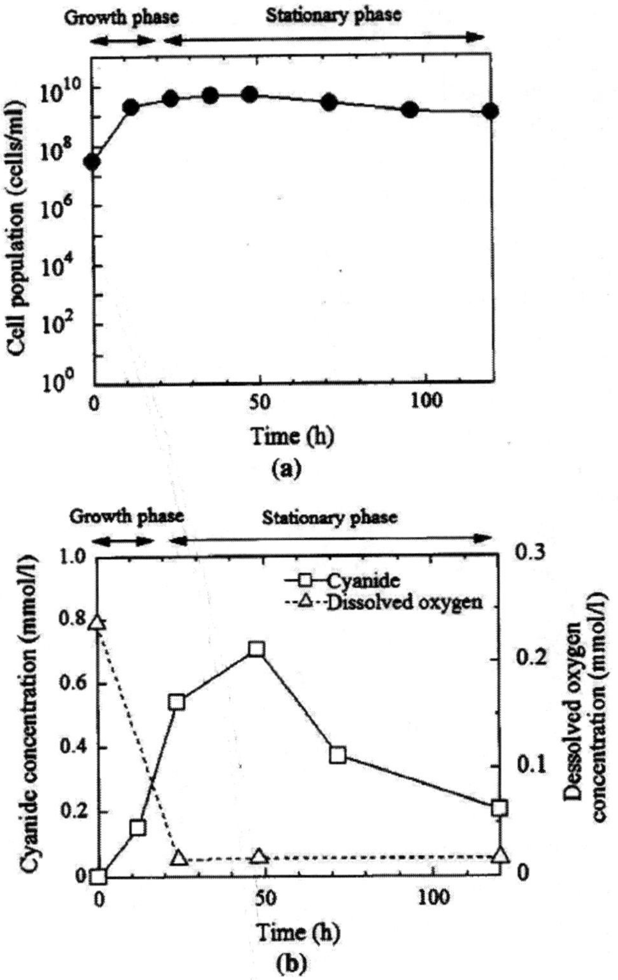

Figure 10.5. Changes of cell population (a), generated cyan concentration and dissolved oxygen concentration in culture medium (b) containing *C. violaceum*.

enhance Au dissolution rate. The additional amount of H_2O_2 is also important because too much addition loses number of bacteria.

The dissolution rate of Au also increases under high cyan concentration in culture. The use of glycine added culture was effective to

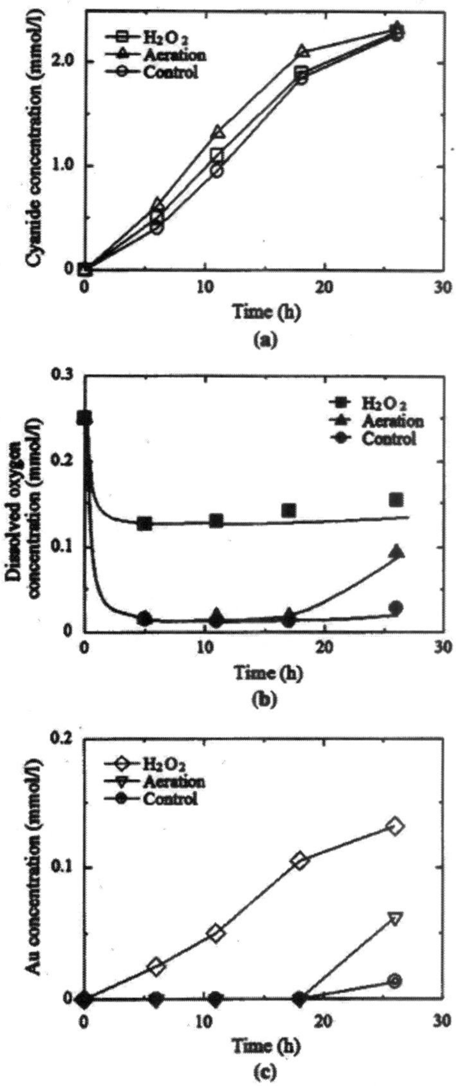

Figure 10.6. Effect of addition of H_2O_2 and aeration on generated cyan concentration (a), change of dissolved oxygen concentration (b), and dissolved Au concentration (c) in *C. violaceum* culture medium.

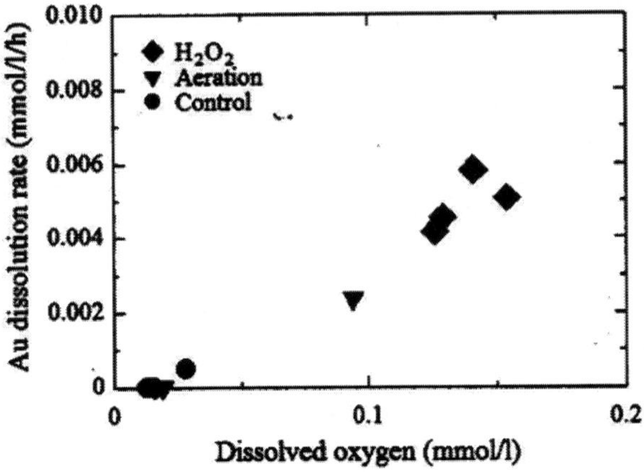

Figure 10.7. Relation between dissolved oxygen concentration in culture medium with *C. violaceum* and dissolution rate of Au.

increase cyan concentration; in this glycine added culture solution it is possible to make solution containing 3 mmol/L cyan.

As mentioned above, it is clear that the dissolution reaction of Au in cyan containing culture proceeded under basic electrochemical reaction. Both increase of cyan generating ability of bacteria and dissolved oxygen concentration in culture medium was confirmed to be effective for enhancement of Au dissolution rate.

10.5 Decomposition of Cyan After Bioleaching

The decrease of cyan concentration was observed after stationary phase during cultivation of bacteria, Fig. 10.2. This means that *C. violaceum* has ability for both generation and decomposition of cyan. The other bacteria such as *Fusarium solani, Fusarium lateritium, Treichoderiuma* etc. are reportedly having ability of cyan decomposition. If generated cyan can be decomposed by bacteria, the environmentally conscious closed loop Au recovery system can be constructed.

Figure 10.8 shows the effect of glucose addition on decomposition of cyan.[19] The test was conducted using glycine added

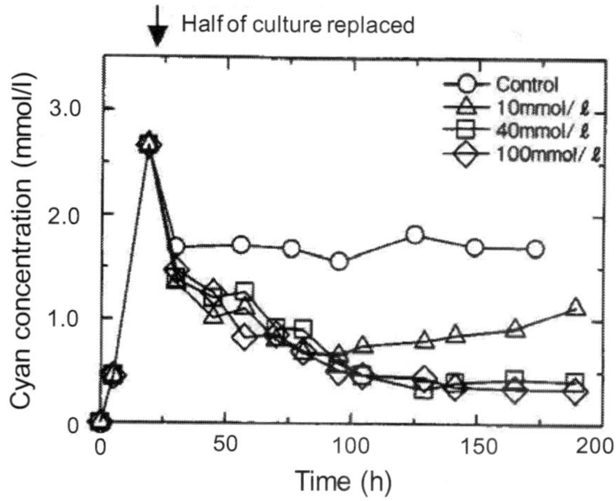

Figure 10.8. Change of cyan concentration after culture for 24 h in glycine added minimum culture and after replacing half of minimum culture with glucose added glutamic acid minimum culture.

YP complex medium that have high ability of cyan generation. Half of culture medium was replaced by glucose added glutamic acid minimal medium at incubation time of 48 h, and then the decrease of cyan concentration was observed. No change of cyan concentration was observed in simple glutamic acid culture medium; however, cyan concentration gradually decreased in glucose added medium. Especially continuous cyan decomposition was succeeded in medium containing more than 40 mmol/L glucose.

The possibility to construct the following closed loop process was indicated by using *C. violaceum*; generation of cyan → dissolution of Au → decomposition of cyan → recovery of Au from detoxified medium → generation of cyan.

The dissolution rate of Au was relatively low under the test condition adopted in this section; however, the cyan containing culture can also dissolve Ni and other metallic elements. Accordingly, the process seems to be effective for recovery valuable metals not only from waste PCBs but also from various Waste Electrical and Electronic Equipment (WEEE).

10.6 Summary

Basic study on bacteria leaching using *C. violaceum* was indicated as an environmentally friendly low impact technology in recovering the precious metals from PWBs. This section described the cyan generating concentration, enhancement of Au dissolution rate, and cyan decomposition under consideration of recovery of precious metals from WEEE. *C. violaceum* gave high cyan generation ability at growth phase in culture; however, the culture became oxygen depleted condition that brought cathode reaction rate controlled process leading the limited low Au dissolution rate. Accordingly, the supply of oxygen into culture was effective to enhance the dissolution rate of Au. However, aeration into culture medium has no effect to enhance Au dissolution at growth phase; at this period the addition of certain amount of H_2O_2 was found to be effective.

To enhance the cyan concentration in culture medium containing *C. violaceum*, the use of glycine added YP culture was effective. By changing half amount of culture medium into glucose added minimum glutamic acid, effective decomposition of generated cyan was achieved. The process seems to be attractive for construction of environmentally conscious recovery system of precious metals from WEEE.

References

1. Fu, B., Zhou, H., Zhang, R., and Qiu, G., *Int. Biodeterior. Biodegradation* **62**, 109–115 (2008).
2. Skladany, G. J. and Metting, Jr., F. B., *Soil Microbial Ecology*, ed. Metting, Jr., F. B. (Marcel Dekker, Inc.) p. 483 (1992).
3. Romero, R., Mazuelos, A., Palencia, I., and Carranza, F., *Hydrometallurgy* **70**, 205–215 (2003).
4. Marhual, N. P., Pradhan, N., Kar, R. N., Sukla, L. B., and Mishra, B. K., *Bioresour. Technol.* **99**, 8331–8336 (2008).
5. Bakhtiari, F., Zivdar, M., Atashi, H., and Seyed Bagheri, S. A., *Hydrometallurgy* **90**, 40–45 (2008).
6. Davis-Belmar, C. S., Nicolle, J. L. C., and Norris, P. R., *Hydrometallurgy* **94**, 144–147 (2008).

7. Zhou, H.-B., Zeng, W.-M., Yang, Z.-F., Xie, Y.-J., and Qiu, G.-Z., *Bioresour. Technol.* **100**, 515–520 (2009).
8. Carranza, F., Romero, R., Mazuelos, A., Iglesias, N., and Forcat, O., *Hydrometallurgy* **97**, 39–45 (2009).
9. Watling, H. R., *Hydrometallurgy* **91**, 70–88 (2008).
10. Frias, C., Diaz, G., Ocana, N., and Lozano, J. I., *Miner. Eng.* **15**, 877–878 (2002).
11. Chan, L. C., Gu, X. Y, and Wong, J. W. C., *Adv. Environ. Res.* **7**, 603–607 (2003).
12. Ren, W.-X., Li, P.-J., Geng, Y., and Li, X.-J., *J. Hazard. Mater.* **167**, 164–169 (2009).
13. Hong, I. T. and Koo, C. H., *Mater. Sci. Eng. A* **393**, 213–222 (2005).
14. Morrison, M. L., Buchanan, R. A., Liaw, P. K., Berry, C. J., Brigmon, R. L., Riester, L., Abernathy, H., Jin, C., and Narayan, R. J., *Diam. Relat. Mater.*, **15**, 138–146 (2006).
15. Zhao, J., Feng, H. J., Tang, H. Q., and Zheng, J. H., *Surf. Coating Tech.* **201**, 5676–5679 (2007).
16. Barrie Johnson, D., Okibe, N., Wakeman, K., and Yajie, L., *Hydrometallurgy* **94**, 42–47 (2008).
17. Xia, L., Liu, X., Zeng, J., Yin, C., Gao, J., Liu, J., and Qiu, G., *Hydrometallurgy* **92**, 95–101 (2008).
18. Lee, M.-S., Ahn, J.-G., and Ahn, J.-W., *Hydrometallurgy* **70**, 23–29 (2003).
19. Kita, Y., Nishikawa, H., Ike, M., and Takemoto, T., *Metall. Mater. Trans. B* **40B**, 39–44 (2009).

Chapter 11

BIOSORPTION OF URANIUM AND/OR THORIUM USING MICROORGANISM

Takehiko Tsuruta

Department of Biotechnology and Environmental Engineering, Hachinohe Institute of Technology, Aza-Ohbiraki 88-1, Myoh, Hachinohe, Aomori 031-8501, Japan

tsuruta@hi-tech.ac.jp

11.1 Introduction

The removal of radionuclide and toxic heavy metals such as uranium and thorium from aqueous solutions, especially from contaminated sources, seems to be a significantly useful subject for environmental control and human health. It is well known that the radiation of radioactive elements, such as uranium, thorium, and lanthanoids, accompanies with nuclear explosion test and the accident of atomic power plants.[1] Fly ash contains considerably high amounts of uranium (30 mgU/kg), and thorium (20 mgTh/kg) is radiated from many thermal power stations.[2] Furthermore, long-term and heavy applications of phosphatic fertilizers mat result in an increase of uranium, thorium, and lanthanoids levels in soils because phosphate rock usually contains relatively high amounts of these elements.[1]

Many researchers have been studying the removal of uranium using microorganisms, such as actinomycetes,[3–5] bacteria,[6–10] fungi,[11–15] and yeasts.[10,16] Thorium removal has also been

Handbook of Metal Biotechnology: Applications for Environmental Conservation and Sustainability

Edited by Michihiko Ike, Mitsuo Yamashita, and Satoshi Soda

www.panstanford.com

investigated by some researchers. Tsezos and Volesky[14,17] reported biosorption of uranium and thorium by some microorganisms and the mechanism of thorium biosorption by *Rhizopus arrhizus*. White and Gadds[15] reported biosorption of thorium by fungal biomass. Andres *et al.*[6] reported adsorption of thorium and uranium by *Mycobacterium smegmatis*. However, little detailed knowledge is available regarding which types of microorganisms can absorb large amounts of thorium.

Recently, I screened various species and strains of actinomycetes, bacteria, fungi, and yeasts for their ability to absorb uranium and/or thorium from the solution containing uranium and/or thorium.

11.2 Screening of the Microorganism to Absorb Uranium and/or Thorium

Various species and strains of actinomycetes, bacteria, fungi, and yeasts were screened for their ability to absorb uranium at pH 5.8[18] and 3.5.[19] Among the microorganisms, a high uranium absorbing ability was exhibited by the Gram-positive bacterial strains, especially *Arthrobacter nicotianae* IAM12342, *Bacillus subtilis* IAM1026, and *Micrococcus luteus* IAM1056 at pH 5.8. The amount of uranium absorbed by Gram-positive bacteria was larger than those by actinomycetes, Gram-negative bacteria, fungi, and yeasts. However, the amount of uranium absorbed using Gram-positive bacteria was strongly affected by the pH of the solution. The amount of uranium absorbed using Gram-positive bacteria was decreased sharply with increasing acidity below pH 5.[18] On the other hand, a high uranium absorbing ability at pH 3.5 was exhibited by the actinomycetes strains, such as *Streptomyces levoris* HUT6156 and *Staphylococcus albus* HUT6047 and Gram-positive bacterial strain, such as *A. nicotianae*. The amounts of uranium absorbed using half strains of actinomycetes were higher than those using all Gram-positive bacterial strains except *A. nicotianae*, all Gram-negative bacteria, fungi, and yeasts.[19]

To determine which type(s) of microorganism(s) has the highest ability to absorb thorium from an aqueous solution, 41 strains of 37 species tested (12 bacteria, 9 actinomycetes, 11 fungi, and 9 yeasts) at pH 3.5 were also screened.[20] Among the microorganisms tested, high thorium biosorption ability was exhibited by certain Gram-positive bacterial strains, notably *A. nicotianae* IAM12342, *Bacillus megaterium* IAM1166, *B. subtilis* IAM1026, *M. luteus* IAM1056, *Nocardia erythropolis* IAM1399, and actinomycetes, especially *S. levoris* HUT6156.

As shown in Fig. 11.1, the amounts of thorium absorbed from the solution containing thorium only[21] or thorium and uranium[22] using most of Gram-positive bacterial strains are higher than those using most of actinomycetes, all of Gram-negative bacteria, fungi, and yeasts strains. The amounts of uranium absorbed from the solution containing thorium and uranium[22] using all microorganisms became lower than those from the solution containing uranium only.[19] However, the amounts of uranium absorbed from the solution containing both elements using half strains of actinomycetes were

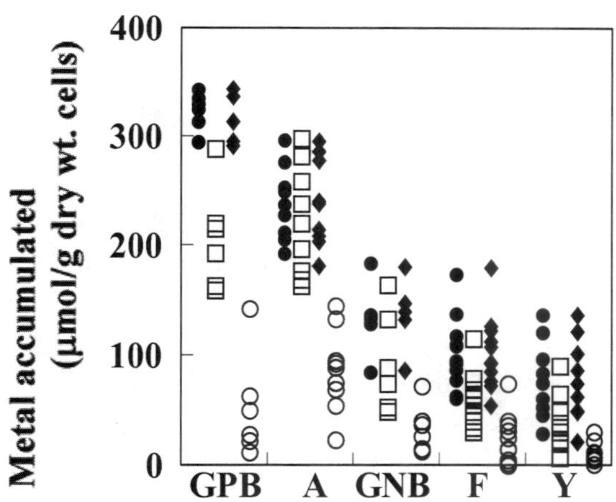

Figure 11.1. Bioaccumulation of uranium and/or thorium using various microorganisms.[23] Resting cells (15 mg dry wt. basis) were suspended in a 100 mL solution (pH 3.5) containing 50 μM uranium and/or thorium for 3 h at 30 °C. GPB: Gram-positive bacteria, A: actinomycetes, GNB: Gram-negative bacteria, F: fungi, Y: yeasts.
Symbols: closed circles: thorium accumulated (μmol/g dry wt. cells) from the solution containing thorium only, squares: uranium accumulated from the solution containing uranium only, diamonds: thorium accumulated from the solution containing uranium and thorium, opened circles: uranium accumulated from the solution containing uranium and thorium.

also higher than those using all Gram-positive bacteria except *A. nicotianae*, all Gram-negative bacteria, fungi, and yeasts.

11.3 The Effect of PH on the Biosorption of Uranium Using *S. Levoris* Cells

Some researchers reported that Gram-negative bacteria,[8,10] fungi,[11,13] and yeasts[10] could absorb uranium from acidic solutions; however, the results in Fig. 11.1 show actinomycetes can absorb larger amounts of uranium than the Gram-negative bacteria, fungi, and yeasts. Generally, the teichoic acid polymers in Gram-positive bacteria confer a strong negative charge on the surface of the cell wall because of their high content of ionized phosphate groups, and little if any teichoic acid is found in Gram-negative bacteria.[24] As a result, it is tentatively considered that chelate formation between the cell surface of Gram-positive bacteria and the uranium becomes stronger than that between the cell surface of Gram-negative bacteria and the uranyl ion. Consequently, it is reasonable that the amount of uranium absorbed by Gram-positive bacteria is larger than that absorbed by Gram-negative bacteria, and is in contrast to the case of gold absorption in which that of gold absorbed by Gram-negative bacteria were found to be larger than that by Gram-positive species.[25]

In our previous studies,[9,18] the amount of uranium absorbed by some bacterial strains decreased sharply below pH 5. Golab *et al.*[5] also reported that the amount of uranium absorbed by *Streptomyces* sp. decreased sharply with increasing acidity below pH 5. However, as shown in Fig. 11.2, the amount of uranium absorbed by *S. levoris* cells was maximum at pH 3.5 to 6.0, but decreased below pH 3.[19] Thus, *S. levoris* cells can absorb uranium from a solution over a wide pH range from 3.5 to 6. It is reasonable to postulate that the amount of uranium absorbed decreases with increasing acidity below pH 3 because the negative charge of the phosphate groups in the teichoic acid polymers of the surface of the Gram-positive bacteria decreases with the increasing acidity of the solution. Additionally, the curve in Fig. 11.2 has a point of inflection between pH 2 and 3. As the pK_{a1} of phosphoric acid is 2.15 (25°C), the pH at which the inflection point

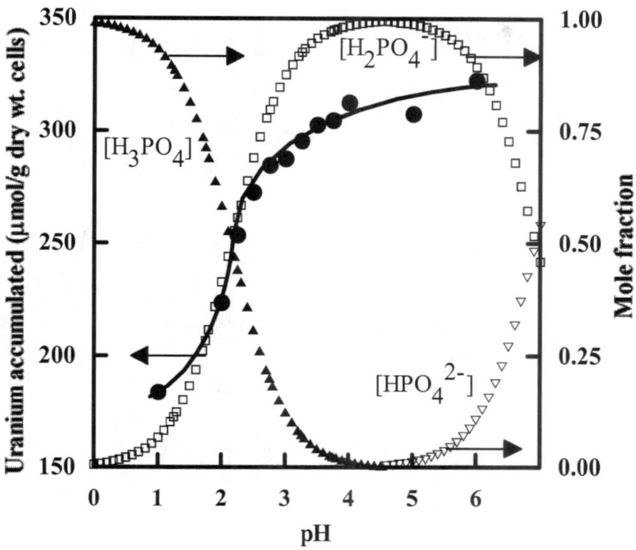

Figure 11.2. Effect of pH on uranium absorption by *S. levoris* cells.[19] Resting cells (15 mg dry wt.) were suspended in a 100 mL solution containing 50 μM uranium for 1 h at 30 °C. Symbols: circles, uranium absorbed (μmol/g dry wt. cells); triangles, mole fraction of H_3PO_4 calculated using K_{a1} of 2.15 (25°C); squares, $H_2PO_4^-$ using the same K_{a1}.

occurs is consistent with the pK_{a1} of phosphoric acid. Accordingly, it is reasonable that the high ability to absorb uranium by *S. levoris* cells at pH 3.5–6.0 is mainly due to the negative charge of the teichoic acid.

11.4 The Effect of External Uranium and Thorium Effect on Their Metals Absorption from the Solution Containing Constant Thorium Concentration and Desired Uranium Concentration Using *S. Levoris* Cells

The effect of the external uranium concentration on the absorption of uranium by *S. levoris* cells was examined.[19] This result is shown in Fig. 11.3. The amount of uranium absorbed by *S. levoris* cells

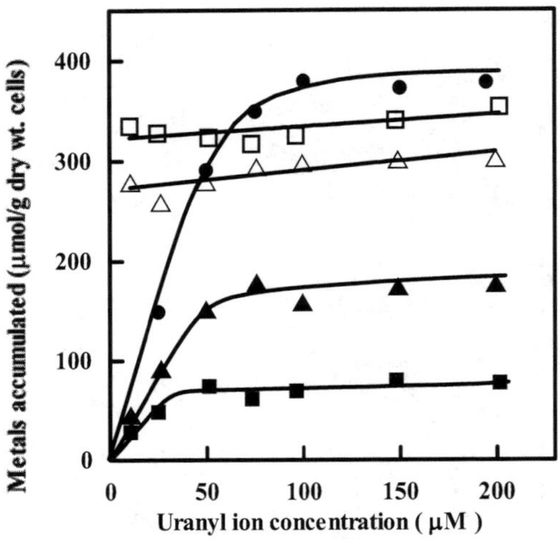

Figure 11.3. Effect of external uranium concentration on uranium (closed symbols) and thorium (open symbols) absorption by *S. levoris* cells.[19] Resting cells (15 mg dry wt.) were suspended in a 100 mL solution (pH 3.5) containing the desired amount of uranium and 0 μM (circles), 50 μM (triangles), and 100 μM (squares) thorium for 3 h at 30 °C.

increased as the external uranium concentration increased. When the external uranium concentration was 100 μM, the maximum uranium absorption, about 380 μmol of uranium per gram dry wt. cells at pH 3.5, was observed.

Another actinoid element, thorium strongly affected uranium absorption.[9,18] As shown in Fig. 11.3, the amount of uranium absorbed by *S. levoris* cells from a solution containing uranium and thorium markedly decreased and that of thorium increased as the external thorium concentration increased during the absorption experiments.

11.5 Time Course of Uranium and Thorium Absorption from the Solution Containing Constant Thorium Concentration and Desired Uranium Concentration Using *S. Levoris* Cells

The time course of uranium absorption by the *S. levoris* cells from a solution containing uranium only was examined.[19] As shown in

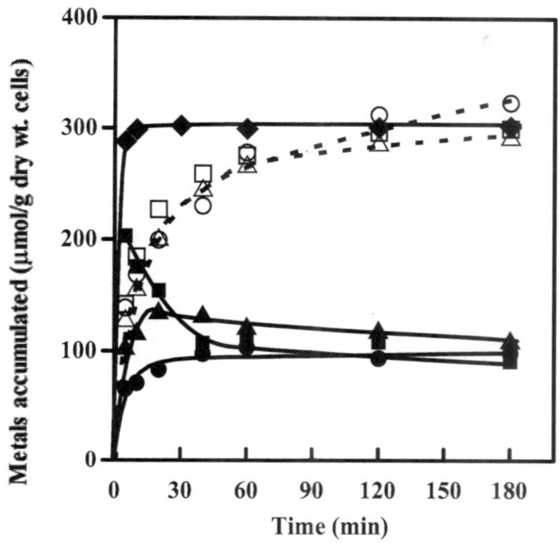

Figure 11.4. Time course of uranium (closed symbols) and thorium (opened symbols) absorption by *S. levoris* cells.[19] Resting cells (15 mg dry wt.) were suspended in a 100 mL solution (pH 3.5) containing 50 μM (diamonds) uranium only (thorium not present), and 50 μM (circles), 75 μM (triangles), and 200 μM (squares) uranium in the presence of 50 μM thorium at 30 °C.

Fig. 11.4, the amount of uranium absorbed by the *S. levoris* cells from a solution containing uranium only increased very rapidly during the first 5 min following the supply of uranium.

However, the time course of the uranium absorption by the *S. levoris* cells from a solution containing uranium and thorium significantly changed. As shown in Fig. 11.4, when uranium absorption from a solution containing 200 μM uranium and 50 μM thorium was monitored, the amount of uranium absorbed by the *S. levoris* cells increased very rapidly during the first 5 min following the addition of uranium, then the uranium absorbed by the *S. levoris* cells desorbed and thorium absorption occurred. These findings indicate that a uranyl–thorium ion exchange reaction occurred at least from a solution containing an excess amount of uranium. When uranium was absorbed from a solution containing 75 μM uranium and 50 μM thorium, the uranyl–thorium ion exchange reaction was also observed. These results indicate that, when excess amounts of uranium are present, uranium is absorbed at first, but uranyl–thorium ion exchange reaction subsequently occurred. Consequently, it is reasonable to conclude that the binding of thorium to the microbial surface is stronger than that of uranium.

11.6 The Effect of External Uranium and Thorium Effect on Their Metals Absorption from the Solution Containing Constant Uranium and Desired Thorium Concentration Using *S. Levoris* Cells

To investigate the ability to absorb thorium by *S. levoris* cells, the absorption test was done from the solution containing the different concentration of thorium.[21] The amount of thorium absorbed was increased with increasing pH of the solution until at pH 3.50.[21] Therefore, the absorption test was done at pH 3.50. As shown in Fig. 11.5, the amount of thorium absorbed by *S. levoris* cells (μmol/g dry wt. cells) increased with increasing external thorium

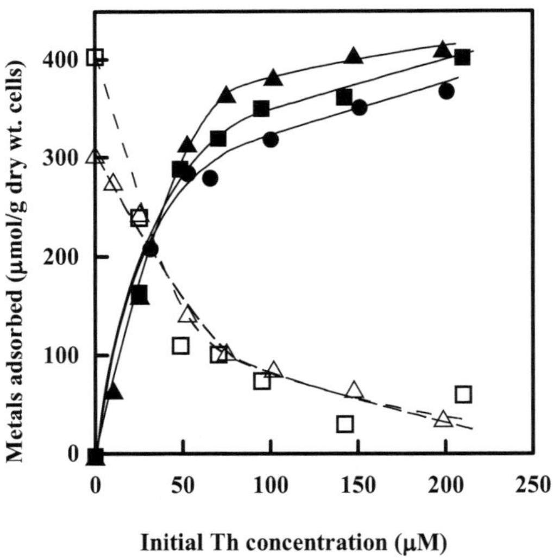

Figure 11.5. Effect of external thorium concentration on thorium (closed) and uranium (opened) absorption by *S. levoris* cells [21]. Resting cells (15.0 mg dry wt. basis) were suspended in a 100 mL solution (pH 3.50) containing a desired amount of thorium and 0 μM (circles), 50 μM (triangles), and 200 μM (squares) of uranium for 3 h at 30 °C.

concentration. When the external thorium concentration was 200 μM, high thorium absorption of about 370 μmol of thorium/g dry wt. cells was observed from the solution containing thorium only. When the effect of external thorium concentration on thorium absorption test was done from the solution containing desired amount of thorium and 50.0 or 200 μM uranium, the amount of thorium absorbed was slightly increased with the addition of uranium, whereas the amount of uranium absorbed decreased with increasing external thorium concentration.

11.7 Effect of Thorium Addition After Uranium Absorption

In the above section, the absorption of thorium was not almost affected by existence of uranium, but the absorption of uranium was strongly affected by existence of thorium. Therefore, to confirm the

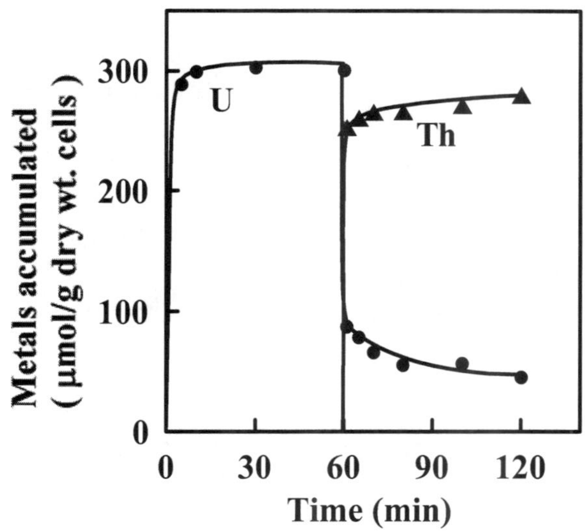

Figure 11.6. Effect of thorium (triangles) addition after uranium (circles) absorption by *S. levoris* cells. Resting cells (15.0 mg dry wt. basis) were suspended in a 100 mL solution (pH 3.50) containing 50.0 μM uranium at 30 °C. At 1 h after from the addition of uranium, the same molar amount of thorium was added to the solution and suspended at 30 °C.

mechanism of the absorption of uranium and thorium, the addition of thorium to the uranium absorbed solution was examined. As shown in Fig. 11.6, the absorption of uranium by the *S. levoris* cells from a solution containing uranium only increased very rapidly. At one hour after the addition of uranium, the same molar amount of thorium was added. After the addition of thorium, the absorbed uranium desorbed and thorium was absorbed rapidly during the first 1 min following the supply of thorium. This fact shows uranium absorbed can be easily desorbed by thorium.

Accordingly, it seems reasonable to have the uranyl–thorium ion exchange reaction occur during the absorption of thorium after the uranium absorption. Therefore, there is a possibility for the occurrence of the uranyl–thorium ion exchange reaction when the absorption was done from the solution containing uranium and thorium.

11.8 Removal and Recovery of Uranium Using Microorganisms Isolated from Uranium Deposits

Additionally, removal and recovery of uranium from aqueous systems using microorganisms isolated from uranium mines were also examined. We identified some strains of bacteria with extremely high uranium removing abilities,[9] and thus, the microbial biomass may be considered as an absorbing agent for the removal and recovery of uranium present in metallurgical effluents, mine tailings, seawater, and other sources.

We have screened hundreds of microorganisms from uranium deposits located in North America, Australia, and Japan for their ability to remove large amounts of uranium. We found several strains which removed large amounts of uranium, including *B. subtilis* from Australia, *Arthrobacter* and *Bacillus* sp. from United States and *Lactobacillus* and *Bacillus* sp. from Japanese uranium deposits.[26] In this study, I also discussed whether the strains of microorganisms isolated can be used for the removal of uranium from uranium refining wastewater and seawater.

11.8.1 *Removal of Uranium from Uranium Refining Wastewater Using Microorganisms Isolated from Uranium Deposits*

The microbial cells with high uranium absorbing ability may be useful for the removal of uranium from uranium mine tailings, uranium refining wastewater, and other waste sources.

As shown in Table 11.1, *Lactobacillus* and *Bacillus* sp. removed 88.1% and 74.4% uranium, respectively, when the pH was only adjusted initially to 6.0.[27] In these studies, the pH value of the solution gradually decreased, with the *Bacillus* cells being more adversely affected by the pH change than the *Lactobacillus* cells. However, both strains quantitatively removed uranium when the pH was continuously adjusted to 6.0. Accordingly, these microorganisms can remove uranium from uranium refining wastewater with a high efficiency.

11.8.2 *Removal of Uranium from Seawater Using Microorganisms Isolated from Uranium Deposits*

The removal of uranium from seawater supplemented with 4.2 μM uranium using the microorganisms isolated from uranium deposits was examined.[27] The amount of uranium removed by *Chlorella* cells from solutions containing 1.196×10^{-3} M sodium hydrogen carbonate was less at pH values above 6 than at pH 5.[28] The decrease in the amount of removed uranium from solutions containing

Table 11.1. Uranium removal from uranium refining wastewater using microbial cells isolated from Japanese uranium mine.

	Removed U (%)	
Strains	pH adjusted only started at pH 6.0	pH adjusted continuously at pH 6.0
Lactobacillus sp.	88.1	99.5
Bacillus sp.	74.4	95.5

Resting cells (15.0 mg dry wt. basis) were suspended in 100 mL of a solution (pH 6.0) of wastewater containing 21.0 μM of uranium for 1 h at 30 °C.

Table 11.2. Removal of uranium using microorganisms isolated from uranium mines .

	Removed U (%)		
Solutions	*Lactobacillus* sp.	*Arthrobacter* sp.	*Bacillus* sp.
Uranium solution (pH 8)	94.7	94.2	94.6
Natural sea water	36.2	0.8	0.9
Decarbonated sea water	70.2	6.1	6.0

Resting cells (15 mg dry wt. basis) were suspended in 100 mL solution (pH 8.0) supplemented with 4.2 μM of uranium for 1 h at 30 °C.

carbonate was estimated from the amount of the UO_2CO_3 formed at pH 6 and of $UO_2(CO_3)_3^{4-}$ formed at pH values greater than 7.[28] As shown in Table 11.2, although *Lactobacillus* sp. removed 36.2% of the uranium from seawater, it removed nearly twice as much (70.2 %) when the seawater was decarbonated.[27] *Arthrobacter* and *Bacillus* cells, which can remove large amounts of uranium from non-saline water, removed far less uranium from either seawater or decarbonated seawater than did *Lactobacillus* cells. Accordingly, the *Lactobacillus* cells have grate potential in applications to remove large amounts of uranium from seawater.

References

1. Tsumura, A. and Yamasaki, S., *Radioisotopes* **42**, 265–272 (1993) (in Japanese).
2. Harmsen, K. and de Haan, F. A. M., *Neth. J. Agric. Sci.* **28**, 40–62 (1980).
3. Byerley, J. J., Scharer, J. M., and Charles, A. M., *Chem. Eng. J.* **36**, B49–B59 (1987).
4. Friiss, N. and Myers-Keith, P., *Biotechnol. Bioeng.* **28**, 21–28 (1986).
5. Golab, Z., Orlowwska, B., and Smith, R. W., *Water Air Soil Pollut.* **60**, 99–106 (1991).
6. Andres, Y., Maccordick, H. J., and Hubert, J.-C., *Appl. Microbiol. Biotechnol.* **39**, 413–417 (1993).
7. Hu, M. Z.-C., Norman, J. M., Faison, B. D., and Reeves, M. E., *Biotechnol. Bioeng.* **51**, 237–247 (1996).
8. Marques, A. M., Roca, X., Simon-Pujol, M. D., Fusto, M. C., and Congregado, F., *Appl. Microbiol. Biotechnol.* **35**, 406–410 (1991).

9. Sakaguchi, T., Tsuruta, T., and Nakajima, A., *Proceedings of the Technical Solutions for Pollution Prevention in the Mining and Mineral Processing Industries*, Palm Coast, FL, USA, 22–27 January 1995; Engineering Foundation Conference, pp. 183–191 (1996).
10. Strandberg, G. W., Shumate, S. E., II, and Parrott, Jr., J. R., *Appl. Environ. Microbiol.* **41**, 237-245 (1981).
11. Galun, M., Keller, P., Malki, D., Fedelstein, H., Galun, E., Siegel, S., and Siegel, B., *Water Air Soil Pollut.* **20**, 221–232 (1983).
12. Galun, M., Keller, P., Malki, D., Feldstein, H., Galun, E., Siegel, S. M., and Siegel, B. Z., *Science* **219**, 285–286 (1983).
13. Treen-Sears, M. E., Volesky, B., and Neufeld, R. J., *Biotechnol. Bioeng.* **26**, 1323–1329 (1984).
14. Tsezos, M. and Volesky, B., *Biotechnol. Bioeng.* **23**, 583–604 (1981).
15. White, C. and Gadds, G. M., *J. Chem. Technol. Biotechnol.* **49**, 331–343 (1990).
16. Shumate, S. E., II, Strandberg, G. W., and Parrott, Jr., J. R., *Biotechnol. Bioeng. Symp.* **8**, 13–20 (1978).
17. Tsezos, M. and Volesky, B., *Biotechnol. Bioeng.* **24**, 955–969 (1982).
18. Tsuruta, T., *J. Biosci. Bioeng.* **94**, 23–28 (2002).
19. Tsuruta, T., *J. Biosci. Bioeng.* **97**, 275–277 (2004).
20. Tsuruta, T., *J. Gen. Appl. Microbiol.* **49**, 215–218 (2003).
21. Tsuruta, T., *Water Air Soil Pollut.* **159**, 35–47 (2004).
22. Tsuruta, T., *J. Nucl. Radiochem. Sci.* **7**, 1–6 (2006).
23. Tsuruta, T., *J. Alloys Comp.* **408–412**, 1312–1315 (2006).
24. Conn, E. E., Stumpf, P. K., Bruening, G., and Doi, R. H., *Outlines of Biochemistry*, 5th Ed. (Wiley, New York, USA) pp. 292–293 (1987).
25. Tsuruta, T., *J. Gen. Appl. Microbiol.* **50**, 221–228 (2004).
26. Sakaguchi, T., Removal of uranium by using microorganisms isolated from Australian and American uraniumdeposits, Environment Innovation in Mining and Mineral Technology, eds. Sanchez, M. A., Vergara, F., Castro, S. H., University of Conception, Chile, pp. 181–191 (1998).
27. Tsuruta, T., *J. Nucl. Sci. Technol.* **43**, 896–902 (2006).
28. Nakajima, A., Horikoshi, T., and Sakaguchi, T., *Agric. Biol. Chem.* **43**, 625 (1979).

Chapter 12

CHARACTERISTICS OF BIOGENOUS IRON OXIDE MICROTUBES FORMED BY IRON-OXIDIZING BACTERIA, *LEPTOTHRIX OCHRACEA*

Jun Takada and Hideki Hashimoto

Department of Applied Chemistry, Graduate School of Natural Science and Technology, Okayama University, Okayama 700-8530, Japan

jtakada@cc.okayama-u.ac.jp

12.1 Introduction

Iron is an abundant and environmentally friendly element. Iron oxides are one of the most important functional materials used as magnetic materials, catalysts, pigments, etc. For example, α-Fe_2O_3, α-FeOOH, γ-Fe_2O_3, and Fe_3O_4 are used as red pigments, combustion catalysts, toner of copying machines, magnetic recording medium of video-tapes, and contrast agents for magnetic resonance imaging, respectively.

Recently, ferrimagnetic ε-Fe_2O_3, one of the polymorphs of Fe_2O_3, was synthesized and its coercivity showed the highest value in metal oxides.[1,2] ε-Fe_2O_3 is expected to be applied to millimeter wave absorbers and intensive researches are now ongoing.[3,4] Because iron

---

Handbook of Metal Biotechnology: Applications for Environmental Conservation and Sustainability
Edited by Michihiko Ike, Mitsuo Yamashita, and Satoshi Soda

www.panstanford.com

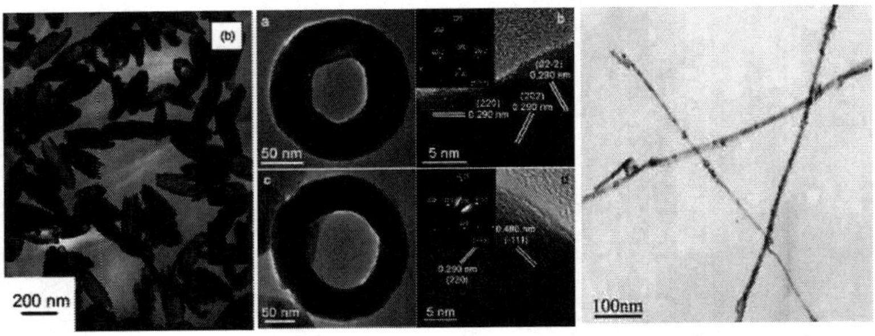

Figure 12.1. Microscope images of synthetic iron oxides. TEM image of the left side shows α-Fe_2O_3 nanotubes.[6] Center TEM image shows Fe_3O_4 nanorings.[7] TEM image of the right side shows Fe_3O_4 nanowires.[8]

oxides are biocompatible materials, magnetic iron oxide nanoparticles (Fe_3O_4 and γ-Fe_2O_3) have received considerable attention in various biomagnetic technologies, such as magnetic fluid hyperthermia for cancer treatment, magnetic separation of cells, magnetic drug targeting, and magnetic resonance imaging.[5]

Nanotechnology researches are also ongoing. Figure 12.1 shows iron oxides having interesting morphologies such as nanotube, nanoring, and nanowire.[6–8] They are prepared by controlling their synthesis conditions strictly. Based on their morphologies and properties, such iron oxides are expected to be applied to templates, microreactors, catalysts, specific magnetic materials, and gas sensors.[6–12]

As just described, iron oxides are extremely important functional materials used in our life and cutting edge researches on iron oxides are now reported. In contrast to such synthetic iron oxides, there are many iron oxides produced by living organisms in nature.

Many living organisms produce iron oxides to use them in their life or as byproduct of their metabolism in nature.[13] For example, limpet and chiton produce α-FeOOH and Fe_3O_4, respectively, to harden their teeth.[13] Iron oxide called ferrihydrite ($FeOOH \cdot 0.4H_2O$)[14] is produced within many living organisms from bacteria to human.[13] Ferrihydrite is the core of ferritin which is an important Fe storage protein. Ferritin controls Fe concentration within living organisms.

Magnetotactic bacteria are very famous bacteria producing magnetic iron oxide. They produce so-called magnetosomes,

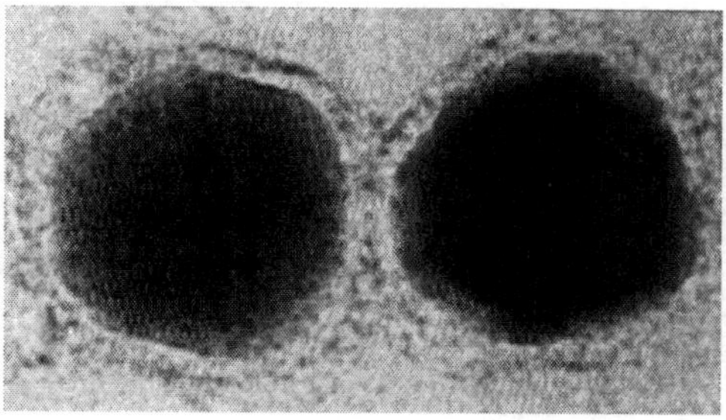

Figure 12.2. Microscope image of magnetosomes produced by magnetotactic bacteria.[17] Magnetosomes are single domain crystalline Fe_3O_4 particles coated with phospholipid bilayer.

Fe_3O_4 nanoparticles coated with phospholipid bilayer, within their cell.[15,16] Figure 12.2 shows transmission electron microscopy (TEM) image of magnetosomes produced by magnetotactic bacteria.[17] Because magnetosomes are single domain crystalline Fe_3O_4 particles showing excellent magnetic properties, intensive researches are currently performed with the aim to use them as magnetic materials in the field of biological and biomedical applications.[18]

Bacteria using Fe^{2+} ions for their metabolism are called iron-oxidizing bacteria. They produce iron oxide precipitates outside their cells by oxidizing Fe^{2+} ions contained in groundwater to Fe^{3+} ions while obtaining necessary energy. However, energy acquisition and Fe oxidation mechanisms have not been clarified yet. It is interesting that iron oxides produced by iron-oxidizing bacteria have extremely unique morphologies such as tubular, twisted stalks-like shapes, etc. The reasons for producing iron oxides with such interesting morphologies also have not been clarified yet.

In the next section we show the characteristics of tubular iron oxides produced by iron-oxidizing bacteria from the point of view of materials science.

12.2 Iron Oxides Produced by Iron-Oxidizing Bacteria

12.2.1 *Iron-Oxidizing Bacteria*

There are many kinds of iron-oxidizing bacteria.[19] They are commonly not so well known but are closely connected with our

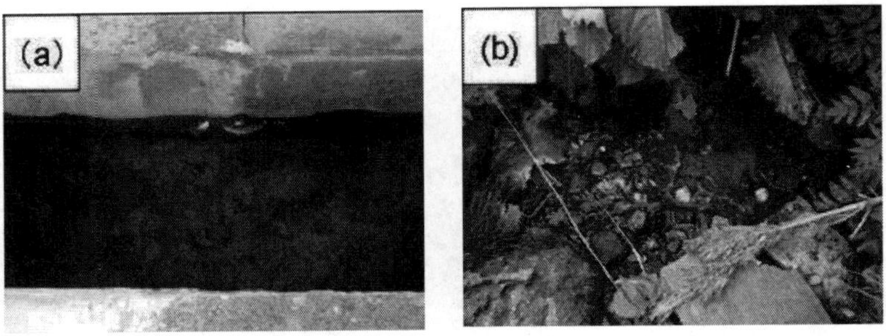

Figure 12.3. Photographs of ocherous precipitates in side ditch. (a) The side ditch in Okayama University in Japan. (b) Beside of small steam in Vitosha mountain in Bulgaria. These ocherous precipitates are colonies of iron-oxidizing bacteria and iron oxides produced by them. See also Color Insert.

everyday life. They are responsible for rusty water damage of tap water, pipe clogging, and ocherous precipitates seen in side ditches, small streams, and springs. Figure 12.3 shows colonies of iron-oxidizing bacteria and ocherous precipitates in side ditch in Okayama University in Japan (Fig. 12.3a) and beside of a stream in Vitosha mountain in Bulgaria (Fig. 12.3b). Such a situation can be observed on many places all over the world.

Figure 12.4 shows microscope images of representative iron oxides produced by iron-oxidizing bacteria.[20–22] Here, we should note that classifications of iron-oxidizing bacteria are based on morphologies of iron oxides produced by iron-oxidizing bacteria and their classification theories are ambiguous and imperfect *Leptothrix ochracea* produces tubular iron oxide with the diameter of approximately 1 μm.[20] *Gallionella ferruginea* produces twisted stalks-like iron oxide[21] and *Toxothrix trichogenes* produces harp-like iron oxide.[22] Iron-oxidizing bacteria produce uniquely shaped iron oxides under ambient temperature and pressure at neutral pH. It is difficult to synthesize such uniquely shaped iron oxides artificially under mild conditions, so that these iron oxides are very interesting from the point of view of materials science.

There are a few iron-oxidizing bacteria which have been isolated successfully. Most of the isolated iron-oxidizing bacteria are members of the genus *Leptothrix*. The genus *Leptotrhrix* is

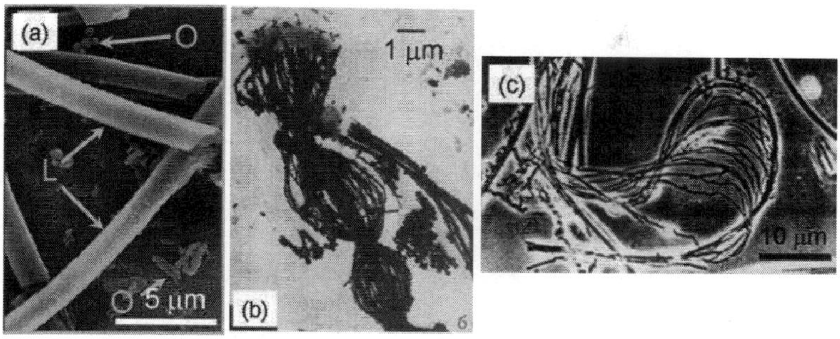

Figure 12.4. Microscope images of representative iron oxides produced by iron-oxidizing bacteria. (a) Tubular iron oxide produced by *Leptothrix ochracea*.[20] (b) Twisted stalks-like iron oxide produced by *Gallionella ferruginea*.[21] (c) Harp-like iron oxide produced by *Toxothrix trichogenes*.[22]

divided into six species, *discophora, cholodinii, mobilis, ochracea, pseudo-ochracea*, and *lopholea*, and isolated species are *discophora, cholodinii*, and *mobilis*.[23] Some strains belonging to other genera (*G. ferruginea* and *Sphaerotilus natans*) have also been isolated.[23,24] However, the other iron-oxidizing bacteria have not been isolated and are only poorly characterized, so there are few studies on iron oxides produced by iron-oxidizing bacteria.

12.2.2 *Water Purification Method by Iron-Oxidizing Bacteria*

Recently, water purification method based on use of iron-oxidizing bacteria is applied in areas where groundwater is abundant in Fe^{2+} ions.[25–28] This method uses ability of iron-oxidizing bacteria to oxidize Fe^{2+} ions contained in groundwater to iron oxides precipitates. *L. ochracea* and *G. ferruginea* are mainly used in this method. This method is simple and low-cost but iron oxide precipitates produced by iron-oxidizing bacteria are unuseful. The picture of water purification plant in Joyo city in Kyoto prefecture in Japan is shown in Fig. 12.5. All ocherous precipitates are iron oxides.

Figure 12.5. Photograph of water purification plant of Joyo city in Kyoto prefecture in Japan. All ocherous precipices are iron oxides produced by iron-oxidizing bacteria. See also Color Insert.

12.2.3 *Characteristics of Biogenous Iron Oxide Microtubes*[29]

In this study iron oxides produced by iron-oxidizing bacteria under ambient temperature and pressure at neutral pH are regarded as a novel functional material and named "biogenous iron oxide." Although there are many kinds of iron-oxidizing bacteria, biogenous iron oxide include all iron oxides produced by iron-oxidizing bacteria.

Especially, we studied biogenous iron oxide produced by *L. ochracea*. The samples were obtained from water purification plant of Joyo city in Kyoto in Japan.

Figure 12.6a shows a typical low magnification scanning electron microscopy (SEM) image of biogenous iron oxide. The sheath had almost straight shape with diameter of approximately 1 μm. The longest sheath had a length of approximately 200 μm. So the sheath had extremely high aspect ratio (length/diameter) of 200. Figure 12.6b shows a high magnification SEM image of the sheath. It is noteworthy that the sheath wall was composed of nanoparticles with diameter of less than 100 nm. To obtain more detailed information on the sheaths, exact diameters of 20 sheaths were measured on SEM images. The average values of inner- and outer-diameter were found to be 1.1 and 1.4 μm, respectively. Thus, the sheath thickness was found to be approximately 0.15 μm.

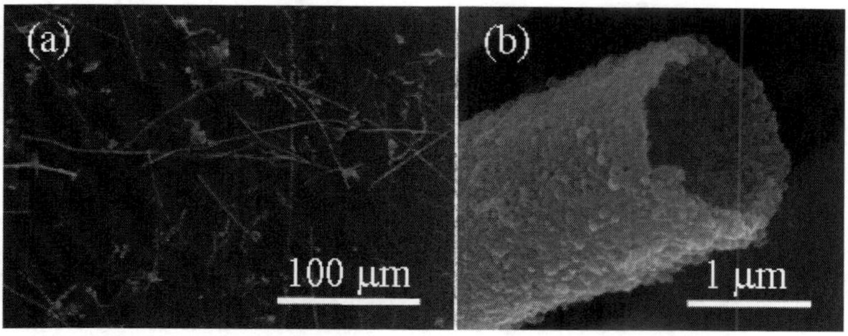

Figure 12.6. Scanning electron microscopy images of biogenous iron oxide produced by *L. ochracea*.[29] (a) Low magnification image. (b) High magnification image.

Such statistic values of the sheath diameters and thickness were first reported in this study. Based on the energy-dispersive X-ray spectroscopy analysis, the chemical composition of the sheaths was roughly determined to be Fe: Si: P = 80: 15: 5 (at %) with the exception of O, H, and C and these elements were distributed uniformly on each sheath.

Figure 12.7 shows a typical XRD pattern of the sheaths. Two broad peaks were observed corresponding to $d = 0.27$ and 0.15 nm. The sheath is considered to be poorly crystalline iron oxide. The XRD pattern is similar to that of two-line ferrihydrite, $Fe_4(O, OH, H_2O)_{12}$.[30]

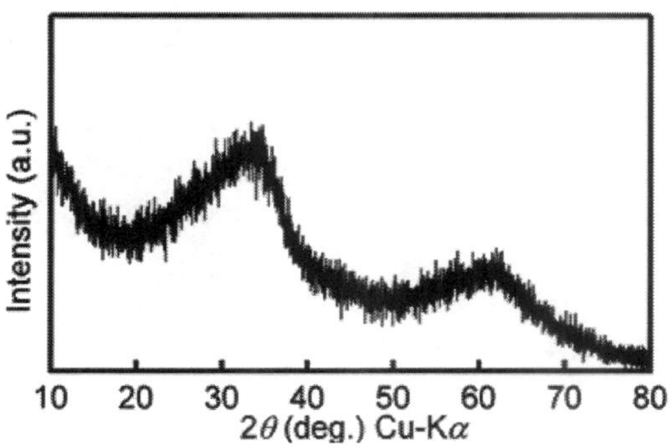

Figure 12.7. Powder X-ray diffraction pattern of biogenous iron oxide produced by *L. ochracea*.[29] Two broad lines showed low crystallinity.

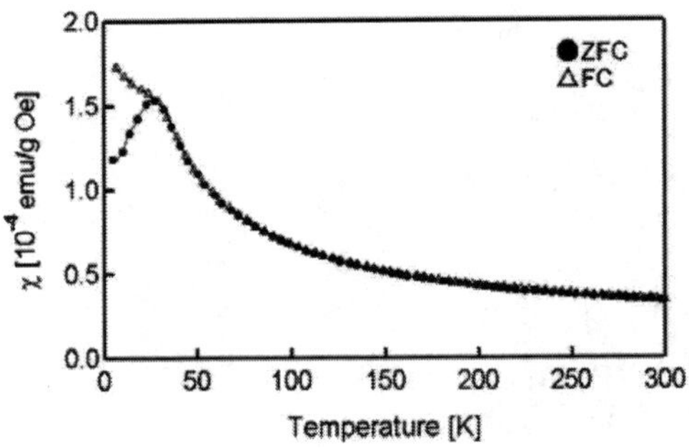

Figure 12.8. The result of direct current susceptibility of biogenous iron oxide.[29] Zero-field cooling (ZFC) and field cooling (FC) of biogenous iron oxide were conducted in the temperature rage from 5 to 300 K under a magnetic field of 1000 Oe.

The temperature dependencies of susceptibility at ZFC and FC curves of the sheaths are shown in Fig. 12.8. The susceptibility curves suggest the spin-glass-like behavior with $T_N = 26$ K.

12.3 Conclusions and Outlook

In this chapter we have shown that biogenous iron oxide produced by iron-oxidizing bacteria are extremely interesting subjects of researches not only in the fields of microbiology and geochemistry but also in the field of materials science. Especially, we introduced the research topic on biogenous iron oxide microtubes produced by *L. ochracea*. From the point of view of materials science, biogenous iron oxide has an interesting morphology, crystallographic structure, microstructure, chemical composition, and magnetic properties. In the near future, we believe that biogenous iron oxide will be applied as a practical functional material in various fields of science.

Acknowledgments

The authors are indebted to Dr. T. Fujii (Okayama University), Dr. M. Nakanishi (Okayama University), Dr. Y. Kusano (Kurashiki University of Science and the Arts), Dr. H. Asaoka (Okayama

University), Dr. Y. Ikeda (Kyoto University), Prof. M. Seno (Okayama University), Dr. R. Murakami (Kyoto National Museum), Prof. H. Kunoh (Okayama University), Dr. T. Suzuki (Okayama University), and Prof. I. Šafařík (Academy of Sciences of the Czech Republic) for helpful discussions and to staff of water purification plant in Joyo city for supplying iron oxide precipitates. This study was partially made by financial support from Special Funds for Education and Research from the Ministry of Education, Culture, Sports, Science and Technology.

References

1. Tronc, E., Chanéac, C., and Jolivet, J. P., *J. Solid State Chem.* **139**, 93–104 (1998).
2. Jin, J., Ohkoshi, S., and Hashimoto, K., *Adv. Mater.* **16**, 48–51 (2004).
3. Ohkoshi, S., *et al.*, *Angew. Chem. Int. Ed.* **46**, 8392–8395 (2007).
4. Namai, A., *et al.*, *J. Am. Chem. Soc.* **131**, 1170–1173 (2009).
5. Maeda, M., Kuroda, C. S., Shimura, T., Tada, M., and Abe, M., *J. Appl. Phys.* **99**, 08H103–08H103-3 (2006).
6. Lv, B., *et al.*, *Particuology* **6**, 334–339 (2008).
7. Jia, C. J., *et al.*, *J. Am. Chem. Soc.* **130**, 16968–16977 (2008).
8. Lian, S., *et al.*, *Solid State Commun.* **132**, 375–378 (2004).
9. Jia, C. J., *et al.*, *Agew. Chem. Int. Ed.* **44**, 4328–4333 (2005).
10. Zhan, S., Chen, D., Jiao, X., and Liu, S., *J. Colloid Interface Sci.* **308**, 265–270 (2007).
11. Liu, Z., *et al.*, *J. Am. Chem. Soc.* **127**, 6–7 (2005).
12. Chen, S., *et al.*, *Mater. Lett.* **59**, 985–988 (2005).
13. Cornell, R. M. and Schwertmann, U., *Iron Oxides*, 2nd Ed. (Wiley-VCH Verlag, Weinheim, Germany) pp. 475–489 (2003).
14. Zhao, J., Huggins, F. E., Feng, Z., and Huffman, G. P., *Clays Clay Miner.* **42**, 737–746 (1994).
15. Blakemore, R., *Science* **190**, 377–379 (1975).
16. Bazylinski, D. A. and Frankel, R. B., *Nat. Rev. Microbiol.* **2**, 217–230 (2004).
17. Schüler, D. *J. Molec. Microbiol. Biotechnol.* **1**, 79–86 (1999).

18. Šafařík, I. and Šafaříková, M., *Manatshefte für Chemie* **133**, 737–759 (2002).
19. Ghiorse, W. C., *Annu. Rev. Microbiol.* **38**, 515–550 (1984).
20. James, R. E. and Ferris, F. G., *Chem. Geol.* **212**, 301–311 (2004).
21. Vatter, A. E. and Wolfe, R. S., *J. Bacteriol.* **72**, 248–252 (1956).
22. Hirsch, P., *The Prokaryotes*, ed. Dworkin, M. (Springer, New York) pp. 986–989 (2006).
23. Spring, S., *The Prokaryotes*, ed. Dworkin, M. (Springer, New York) pp. 758–777 (2006).
24. Hanert, H. H., *The Prokaryotes*, ed. Dworkin, M. (Springer, New York) pp. 990–995 (2006).
25. Mouchet, P., *J. Am. Water Works Assoc.* **84**, 158–167 (1992).
26. Katsoyiannis, I. A. and Zouboulis, A. I., *Water Res.* **38**, 1922–1932 (2004).
27. Søgaard, E. G., Medenwaldt, R., and Abraham-Peskir, J. V., *Water Res.* **34**, 2675–2682 (2000).
28. Søgaard, E. G., Aruna, R., Abraham-Peskir, J. V., and Koch, C. B., *Appl. Geochem.* **16**, 1129–1137 (2001).
29. Hashimoto, H., *et al.*, *J. Magn. Magn. Mater.* **310**, 2405–2407 (2007).
30. Eggleton, R. A. and Fitzpatrick, R. W., *Clays Clay Miner.* **36**, 111–124 (1988).

Chapter 13

MICROBIAL FORMATION OF SEMICONDUCTOR NANO-PARTICLES CONTAINED SELENIUM OR TELLURIUM AND METALS

Toshifumi Sakaguchi
Department of Environmental Sciences, Prefectural University of Hiroshima, 562, Nanatsuka, Shobara, Hiroshima, 727-0023, Japan
sakaguchi@pu-hiroshjima.ac.jp

13.1 Introduction

Microbial formation of elemental selenium (Se) or tellurium (Te) has already been reported by 1960s.[1–4] Depositions of Se or Te granule have been observed in many species of microorganisms. The intra- and extracellular formations of Se or Te fine nano-particles had been attributed to the reduction of Se or Te oxyanions due to detoxification mechanism[5–7] or retention of the electron poise[8–11] in microbes. Furthermore, recent report disclosed that those nano-particles generate it as results that Se oxyanions were reduced to elemental Se as an electron acceptor for growth in the anaerobic

Handbook of Metal Biotechnology: Applications for Environmental Conservation and Sustainability
Edited by Michihiko Ike, Mitsuo Yamashita, and Satoshi Soda

www.panstanford.com

respiration.[7,12,13] Similar possibility is also suggested about Te oxyanion and its anaerobic respiratory metabolism.[14] Many microorganisms are involved in the geobiochemical cycles of Se and Te due to their abilities in enzymatic reduction.[15–22] Cosmopolitan distribution of microbes capable of reducing Se or Te oxyanions in various environments suggests that elemental Se and Te crystals can be formed in diverse natural sediments and soils.[23,24] So far, many chemical methods for neutralizing or cohesive precipitation have been used to remove poisonous substances such as heavy metals and metalloid elements from mining, refining, manufacturing, and power generation drains. However, these chemical treating methods have to administer a large amount of reagents to drain. In addition, it is necessary to remove the administered chemical treatment reagents. Although heavy metals and metalloid elements can be essential to industrial material production as rare element, the effective recovery method of these released or contaminated elements have not been developed. Further research should explore the biological function which can apply to develop the recovery and recycling technology. Nowadays, metal and metalloid removals by bacterial reduction represent a bioremediation process.[8–11,19,24–27]

First report on microbial production of quantum dots semiconductor nano-particles is due to formation of nanometer-scale cadmium sulfide (CdS) crystallites that were produced in yeasts *Candida glabrata* and *Schizosaccharomyces pombe* by their detoxification processes related to metallothionine.[28] Since this research was reported, further researches on the mass production and the application are performed. Production of biogenic semiconductor CdS nano-particles has been achieved in batch culture of *S. pombe*.[29] The isolation and partial characterization of a highly Cd-resistant *Klebsiella planticola* strain Cd-1 capable of anaerobically producing CdS at high cadmium concentration (15 mM $CdCl_2$) have been reported.[30] Such microbial isolate capable of converting cadmium ion to CdS precipitates is expected as a biocatalyst for the accelerated bioremediation of systems contaminated by high levels of cadmium. Moreover, X-ray scattering analysis of the nano-particles and their use in the fabrication of an ideal diode were carried out.[31] Their data showed that the biogenic CdS was a particle in the range of 1 to 1.5 nm, which had a Wurtzite ($Cd_{16}S_{20}$)-type hexagonal lattice structure. The constructed diode exhibited less than 75 mA/cm^2 current at 10 V when forward biased and the breakdown occurred at less than 15 V in the reverse biased mode. The results showed that the feature can be regarded as an ideal diode.[31] Biogenic semiconductor has been attempted to fabricate electronics devices using II–VI semiconductor materials. Recently, microbial manufacture of Se-based nano-particles by the reduction of Se oxyanion with a selenite-reducing microorganism, *Veillonella atypica*, has been reported.[32] The ability of metal-reducing microorganisms to synthesize nano-particles can be used for the biomanufacture of fluorescent and semiconducting nano-materials. Thus, in this chapter, we would like to introduce our attempts and performances on the simultaneous removals of Te oxyanions (e.g. tellurite) or Se oxyanions (e.g. selenate) and heavy metal cations (e.g. cadmium chloride) due to conversion to Se or Te and metal containing semiconductor nano-particles (e.g. cadmium telluride: CdTe) by using anaerobic reactions of microbial consortia and their isolates.

13.2 Isolation and Enrichment

Many sludge and sediment samples were collected from various terrestrial, freshwater and marine environments such as polluted urban rivers, estuary, and oil-producing wells or ponds (Fig. 13.1). These samples were inoculated to the anaerobic isolation medium. For isolation and enrichment of microorganisms the isolation medium contained was newly designed by the modification of the growth medium for a selenate-reducing bacterium *Citrobacter* sp. strain JSA.[33,34] Microorganisms were enriched and incubated in the isolation medium contained at 1 mM of Se or Te oxyanion (e.g. sodium selenate) and heavy metal cation (e. g. cadmium chloride), respectively, under anaerobic conditions at 25°C. Black or red precipitates that considered to be corresponding to semiconducting nano-particles were formed in the enrichment. In particular, the change was prominent in the enrichments which were inoculated samples from oil-producing wells and ponds (Figs. 13.2 and 13.3). Elemental analyses of heavy fuel oil[35] and crude oil[36] revealed that trace metals such as Ni, V, Cu, Cd, Fe, Mg, and Pb were detected as

Figure 13.1. Natural oil-producing wells and ponds in Japan, where samples were collected for the isolation and enrichment of microorganisms. a: Oil-producing wells in Niitsu, Niigata; b: a natural oil-producing pond in Singleton memorial park (Abura-tsubo: Oil vessel); c: a natural oil-producing pond in Showa, Akita; d, e, f: oil-producing wells in Showa and Akita city, Akita. See also Color Insert.

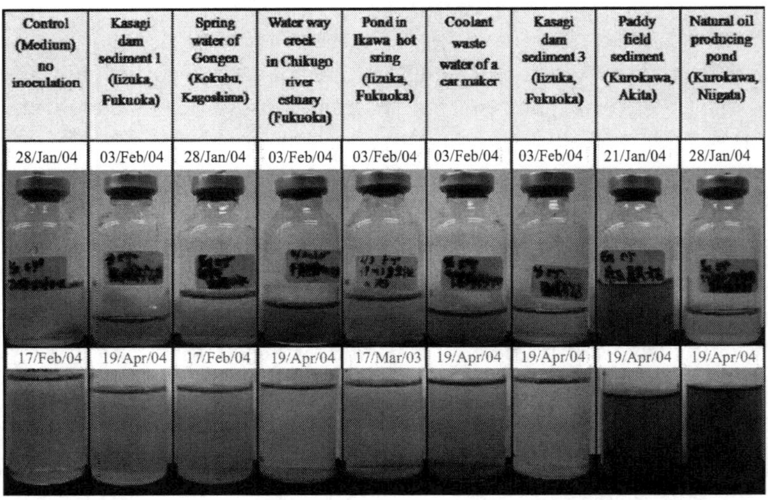

Figure 13.2. Appearance of precipitation in enrichment contained zinc and selenate (1 mM respectively) for ZnSe formation (upper line: sampling place, middle line: 1st enrichment, lower line: transferred enrichment (>2nd)). See also Color Insert.

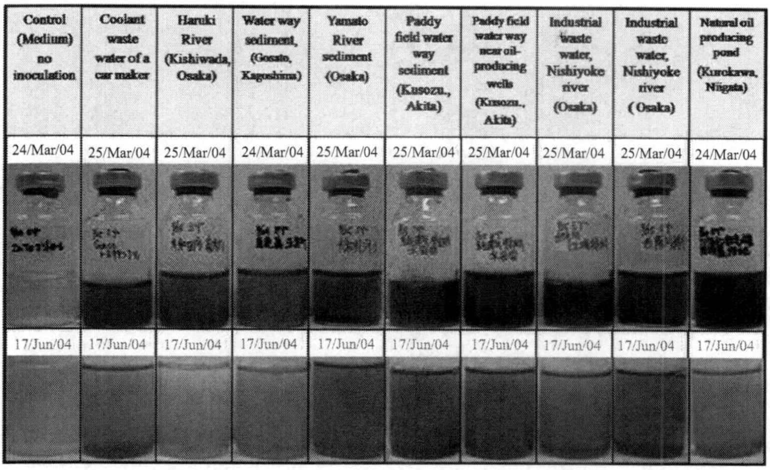

Figure 13.3. Appearance of precipitation in enrichment contained zinc and tellurite (1 mM respectively) for ZnTe formation (upper line: sampling place, middle line: 1st enrichment, lower line: transferred enrichment (>2nd)). See also Color Insert.

normal constituent of their mineral components. Such place is an environment where the presence of metal-resistant or -reducing microorganisms is strongly expected. It may be the major cause of biogenic precipitations in their enrichments under the condition including heavy metal and Se or Te oxyanions. In our research, red or black precipitates markedly arose in a number of enrichments contained zinc and selenate ions (for ZnSe; Fig. 13.2), zinc and tellurite ions (for ZnTe; Fig. 13.3), and cadmium and tellurite ions (for CdTe; Fig. 13.4). They were inoculated samples which come from oil-producing wells or ponds in Akita and Niigata. Enrichments capable of producing precipitates were selected when transferred to the fresh medium. Microbes were isolated by the colony formation method in the anaerobic conditions from their enrichments. Isolated microbial strains and stable consortia were maintained anaerobically. As a result of this experimental procedure, we have succeeded to obtain consortia (NT-ER and SM-ER) and pure isolates (strains SM-8 and 9) capable of synthesizing CdTe nano-particles from enrichments (originated from a natural oil-producing pond or

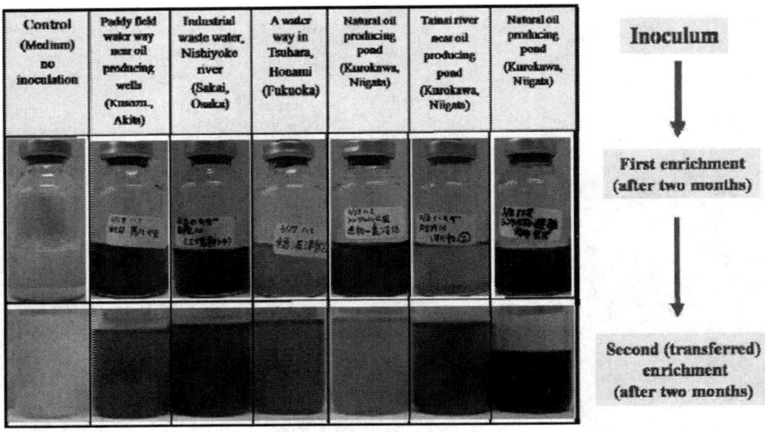

Figure 13.4. Appearance of precipitation in enrichments contained cadmium and tellurite (1 mM respectively) for CdTe formation. See also Color Insert.

river sediment in/near Singleton memorial park, Kurokawa, Tainai, Niigata) in the presence of cadmium and tellurite ions (Fig. 13.4) . In addition, we have also established numerous consortia and isolates that can form precipitates which are considered to be nano-fine particles such as CdSe, CoSe, NiSe, ZnSe, CdSe, CoTe, NiTe, and ZnTe.

13.3 Extraction of Nano-Particles

After cultivation and centrifugation of culture, ultra sonic or enzymatic disruption of the harvested cells from enrichment was carried out. Both techniques of ultra sonication and enzymatic treatment were effective for the extraction of nano-particles. Especially, proteinase K treatment at 55°C for 12 hours was more efficient in harvesting only nano-particles from cells. The collected nano-particles were washed and rinsed by distilled water for the elemental analysis and transmission electron microscopy (TEM) observation. The size could be drawn by filtration technique with a cellulose nitrate membrane filter. Nano-particles less than 200 nm were prepared by this method.

13.4 Elemental and Crystal Analysis of Nano-Particles

After disruption of harvested cells, the extracted particles were analyzed with energy dispersive X-ray microanalysis (EDX) and high-resolution transmission electron microscopic observation (HRTEM). In case of particles from the enrichment (SM-ER) for CdTe production, the analysis showed the microbial depositions were nano-sized particles that were composed of Cd and Te with approximately 10 nm in diameter (Fig. 13.5a,b). Furthermore, Cd and Se were similarly detected by the elemental analysis in case of nano-particles isolated from the enrichment for CdSe production (Fig. 13.6a). The purified particles were almost within approximately 10–20 nm in diameter (Fig. 13.6a,b). The extracted particles had the small globular shapes that differ from the biogenic crystals of elemental Se with smooth morphologies (Fig. 13.6b). As shown in Fig. 13.5, the formation of aggregates which composed of particles with approximately 10 nm diameter was observed. Morphology of these

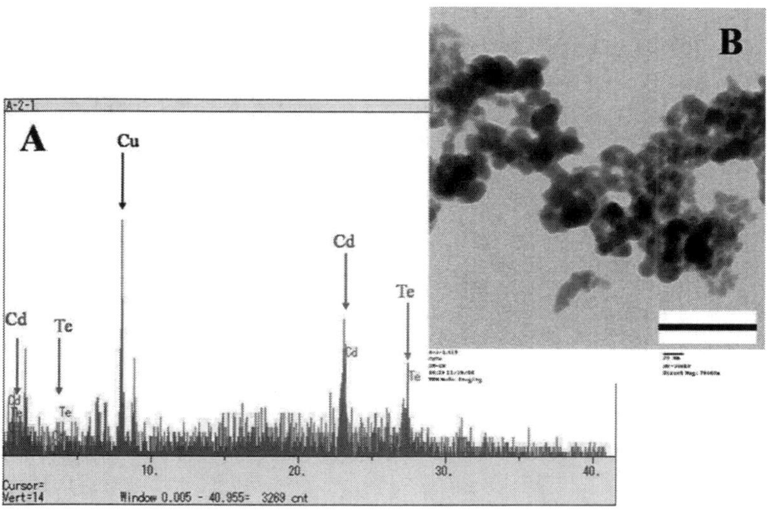

Figure 13.5. Energy dispersive X-ray analysis (a) and HRTEM observation of biogenic nano-particles in the enrichment (SM-ER) contained cadmium and tellurite ions (1 mM respectively) for the formation of CdTe Cu signals are from a mesh grid (a). Bar indicates 100 nm (b). Hitachi H-9000NAR was used for the EDX analysis and HRTEM observation.

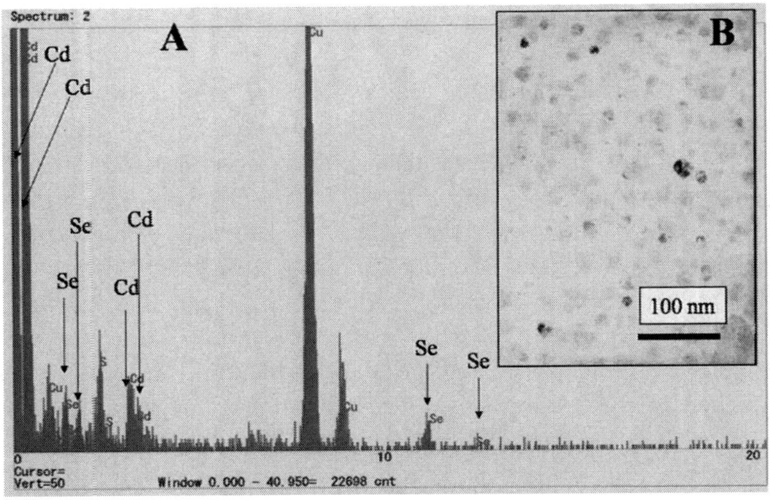

Figure 13.6. Energy dispersive X-ray analysis (a) and HRTEM observation of biogenic nano-particles in the enrichment contained cadmium and selenate ions (1 mM respectively) for CdSe formation Cu signals are from a mesh grid (a). Bar indicates 100 nm (b). Hitachi H-9000NAR was used for the EDX analysis and HRTEM observation.

particles was completely different from the needle-like shape of biogenic elemental Te crystals (Fig. 13.5b). Results on the HRTEM observation and EDX analysis strongly supported that the produced particles were CdTe or CdSe. Furthermore, similar results were obtained in the case of biogenic nano-particles in the enrichment contained zinc and tellurite ions for the formation of ZnTe (Fig. 13.7). Furthermore, electron diffraction analyses on produced particles in their enrichments indicated that they contained crystalloid structure (Fig. 13.8). Electron diffraction image (Debye–Scherrer ring) of biogenic particles in enrichment (NT-ER) and the lattice interplanar spacing in each Miller index were consistent with that of standard CdTe (sphalerite). This result demonstrated microbial formation of CdTe with sphalerite (zicblende) structure. Likewise, formations of CoSe, NiSe, ZnSe, CdSe, CoTe, NiTe, and ZnTe in their cultures that coexisted with Se or Te oxyanion and heavy metal ion were revealed in similar manner as describe above.

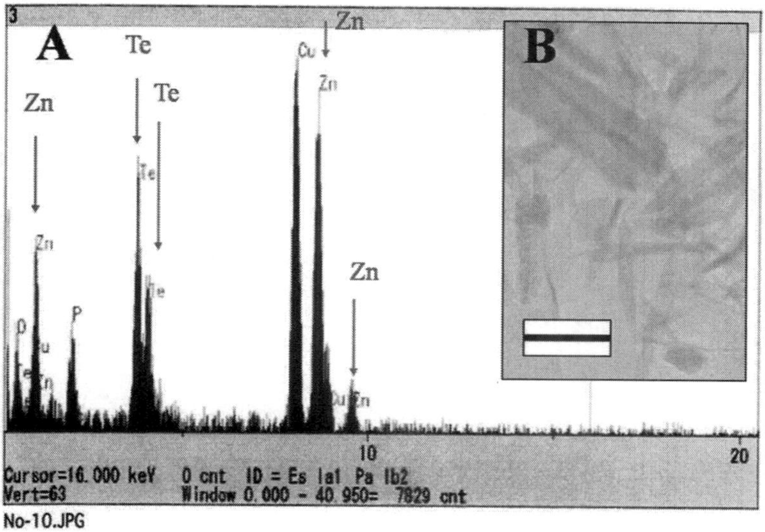

Figure 13.7. Energy dispersive X-ray analysis (a) and HRTEM observation of biogenic nano-particles in the enrichment contained zinc and tellurite ions (1 mM respectively) for ZnTe formation Cu signals are from a mesh grid (a). Bar indicates 100 nm (b). Hitachi H-9000NAR was used for the EDX analysis and HRTEM observation.

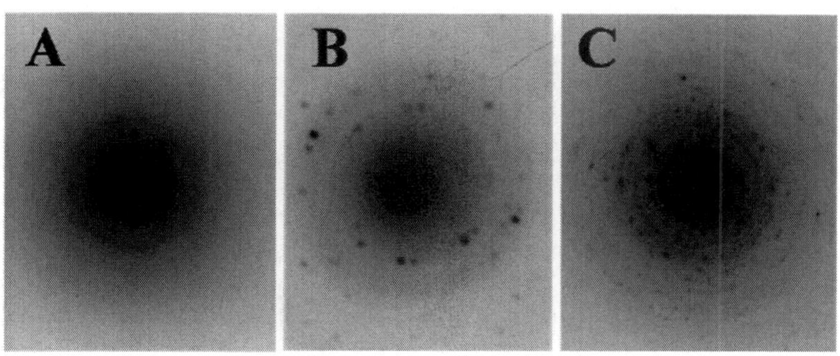

Figure 13.8. Electron diffraction images (Debye–Scherrer ring) of biogenic nano-particles. a: candidate of biogenic CdTe particles, b: candidate of biogenic CdSe particles, c: candidate of biogenic ZnTe particles (camera length: 1.1 m, Electron beam: 100 kV, HRTEM: Hitachi H-9000NAR).

13.5 Isolation of Microbe and Denaturing Gradient Gel Electrophoresis Analysis in Enrichments

We have obtained two pure strains of isolates from enrichments (SM-ER and NT-ER consortia), which can sustain in the presence of 1 mM of potassium tellurite and cadmium chloride, respectively. Three microbial isolates, designated strains NTT-1, 2, and 3, respectively, were isolated from the enrichment NT-ER that can produce CdTe particles (Fig. 13.9: a1 in case of strain NTT-1). Moreover, two strains of bacteria, designated strains SM-8 and 9, were obtained from the enrichment (SM-ER) capable of producing CdTe nano-particles (Fig. 13.9: b1 in the case of strain SM-8). All isolates from NT-ER and SM-ER (consortia) enrichments were gram-negative short-rods or cocci with the size of 1–3 μm. Elemental Te crystals

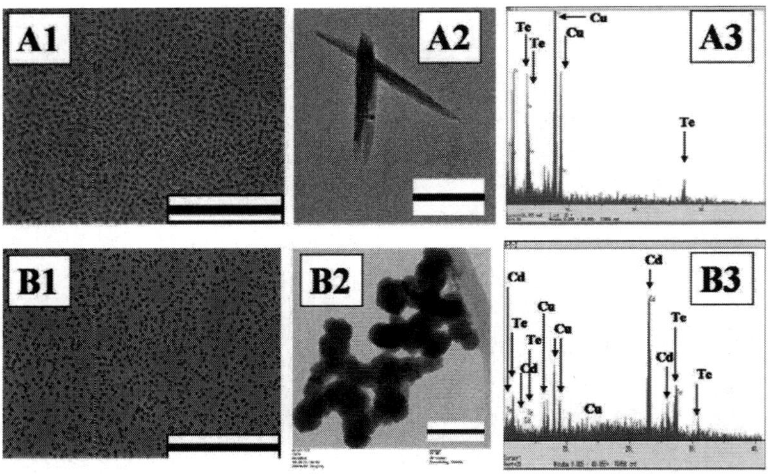

Figure 13.9. Phase-contrast photomicrographs of microbial isolate from enrichments (NT-ER and SM-ER consortia), and TEM image and EDX elemental analysis of the nano-particle produced by the isolate. a1 and b1 show phase-contrast photomicrographs of strains NTT-1 (from consortium NT-ER) and SM-8 (from consortium SM-ER), respectively. a2 and a3 indicate TEM image and EDX analysis of the produced nano-particles by strain NTT-1, respectively. b2 and b3 indicate TEM image and EDX analysis of the produced nano-particles by strain SM-8, respectively. Bars indicate 50 μm (in a1 and b1) and 100 nm (in a2 and b2). Cu signals come from the mesh grid in a3 and b3. See also Color Insert.

Figure 13.10. PCR-DGGE analysis of microbial consortium SM-ER and the microbial isolates, strains SM-8 and SM-9. lane 1: SM-ER consortium incubated for one month, lane 2: SM-ER consortium incubated for two months, lane 3: strain SM-8 incubated for one month, lane 4: strain SM-9 incubated for one month. Forward primer (5′-GC clamp-CCTACGGGAGGCAGCAG-3′) corresponding to *E. coli* numbering position 341–357 (Bacteria V3 region), reverse primer (5′-CCGTCAATTCCTTTAAGTTT-3′) corresponding to *E. coli* numbering position 907–926 (Universal V5 region). 6% polyacrylamide gel (20%–80% gradient denaturing solution) was used for the electrophoresis at 200 V for four hours. See also Color Insert.

with characteristic needle-like shape were produced in the culture of strains NTT-1, 2, and 3, respectively (Fig. 13.9: a2 and a3 in the case of strain NTT-1). CdTe nano-particles were not found in their cultures, whereas strains SM-8 and 9 could synthesize CdTe nano-particles in the cultures, respectively (Fig. 13.9: b2 and b3 in the case of strain SM-8). CdTe could be produced by culturing strain SM-8 or 9 alone. Furthermore, PCR-DGGE (denaturing gradient gel electrophoresis) analysis of microbial consortium SM-ER and the isolates (SM-8 and SM-9) suggested that the microbial community predominantly consisted of two microbial groups corresponding to strain SM-8 and SM-9, respectively (Fig. 13.10). Component of consortium SM-ER was almost constant during the incubation for two months (Fig. 13.10). This result suggested that microbial formation of CdTe in the consortium SM-ER was already caused by strains SM-8 and SM-9 at the initial stage of the incubation. Meanwhile, in consortium NT-ER, the fact that all isolates such as strain NTT-1 cannot result in the formation of CdTe particles alone suggested that symbiosis with the other microbes and/or the component change of the microbial community are required for the CdTe formation in the consortium.

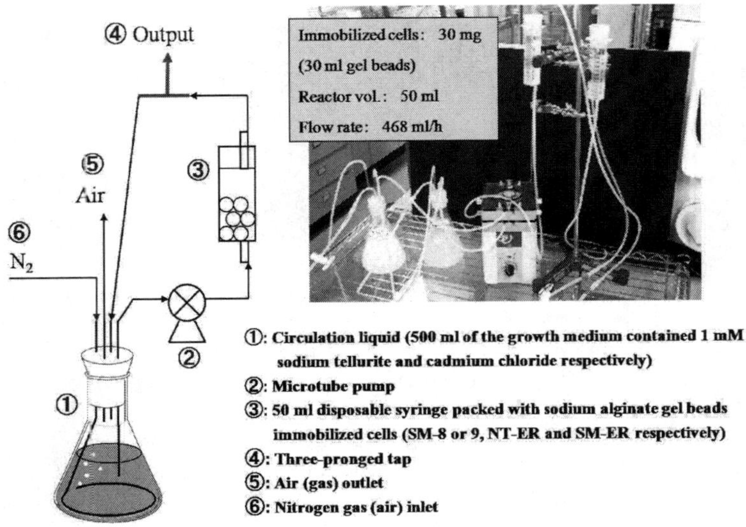

Figure 13.11. Bioreactor system for the simultaneous removal of Cd and Te ions based on conversion to CdTe particles. See also Color Insert.

13.6 Recovery and Conversion of Heavy Metals and Te

Finally, we have attempted the simultaneous removal of Cd (cadmium ion) and Te (telluride) based on conversion to CdTe precipitates by a circular-type bioreactor system with sodium alginate gel beads which immobilized cells of the consortium (SM-ER or NT-ER) or the isolate (strain SM-8 or SM-9). The bioreactor system was designed and constructed as shown in Fig. 13.11. A disposable syringe (50 mL) packed with sodium alginate gel beads that were immobilized cells was prepared and flowed the circulation liquid (500 mL of the growth medium) which contained 1 mM sodium tellurite and cadmium chloride, respectively. The bioreactor was operated for seven days at 25°C under anaerobic condition. Tellurite concentration in the circulation liquid was determined colorimetrically with diethyldithiocarbamate by measuring absorbance at 340 nm.[37] Moreover, the cadmium ion was measured by atomic absorption spectrophotometry (Shimazu, AA-6300). Appearance of the cell-immobilized sodium alginate gel beads in the bioreactor has

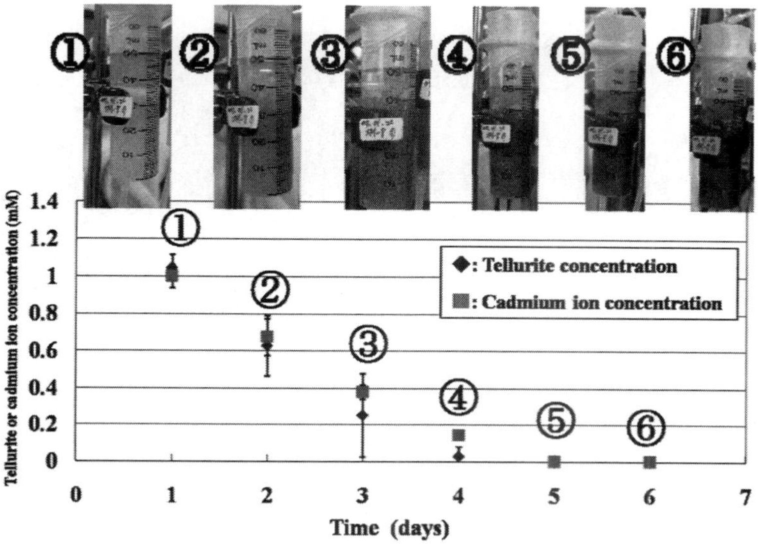

Figure 13.12. Appearance of bioreactor packed with sodium alginate gel beads that immobilized cells of the isolate (strain SM-8), and concentrations of tellurite and cadmium ion in the circulation liquid of the bioreactor system. See also Color Insert.

turned blackish with progress of the running time (Fig. 13.12: in the case of strain SM-8). Tellurite (1 mM) completely disappeared from the circulation solution of the bioreactor in about five days, coincident with the decrease of cadmium ion concentration. The TEM observation, EDX analysis, and the electron diffraction images (Debye–Scherrer ring) represented that both ions could be completely recovered and converted into CdTe precipitates (Fig. 13.13). Likewise, these results were obtained in the cases of strain SM-9 and consortia NT-ER and SM-ER cells as well as strain SM-8. This fact demonstrated the microbial activity could be applied to the simultaneous recovery of Cd and Te with bioreactor system and the conversion to useful nano-particles (CdTe) of harmful heavy metal and toxic anionic Te metalloid. Our results on this attempt implied that the ability of metal-reducing microorganisms to synthesize nano-particles could be used for the biomanufacture of fluorescent and semiconducting nano-materials as well as bioremediation and biorecovery.

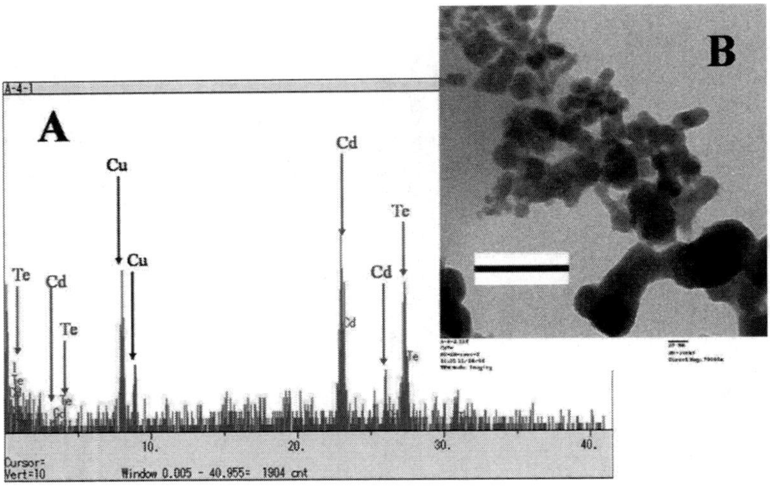

Figure 13.13. Energy dispersive X-ray analysis (a) and HRTEM (b) observation of biogenic nano-particles that were produced in the bioreactor with sodium alginate gel beads which were immobilized cells of consortium NT-ER. Bar indicates 100 nm in b. Cu signals are from a mesh grid (a). Bar indicates 100 nm (b). Hitachi H-9000NAR was used for the EDX analysis and HRTEM observation.

13.7 Conclusions

In this chapter, we have introduced the isolation and enrichment of microorganisms capable of producing nano-particles that can be used for fluorescent and semiconducting devices. Our research revealed that microbial isolates and consortia that were obtained from various environmental samples, such as freshwater, marine, mining water, and spring water sediments and natural oil-producing wells, have the ability to synthesize semiconducting selenide or telluride nano-particles like CoSe, NiSe, ZnSe, CdSe, CoTe, NiTe, ZnTe, and CdTe in their cultures or enrichments in which Se or Te oxyanion and heavy metal cation coexists. These facts show that the microbial function can be applied to not only bioremediation of heavy metals and Se or Te oxyanion pollutions and the simultaneous recovery of harmful Se or Te metalloid (eg. selenite, selenate, and tellurite) and toxic heavy metals such as Ni, Co, Zn, and Cd but also biomanufacture of quantum-dots semiconducting materials. Especially, in the case of CdTe, we have demonstrated that it is possible to recover simultaneously cadmium ion and tellurite based on conversion to CdTe nano-particles valuable of material industry by using bioreactor system. Finally, with respect to the produced (recovered) particles, UV irradiation (365 mm) to the suspension strongly suggests the biogenic particles may be fluorescent to UV (at 365 nm) excitation (Fig. 13.14). At present, we are progressing on the further analysis and evaluation as semiconducting material of the biogenic particles. In future, if the good evaluation was confirmed by our research, our found or established microorganisms and consortia

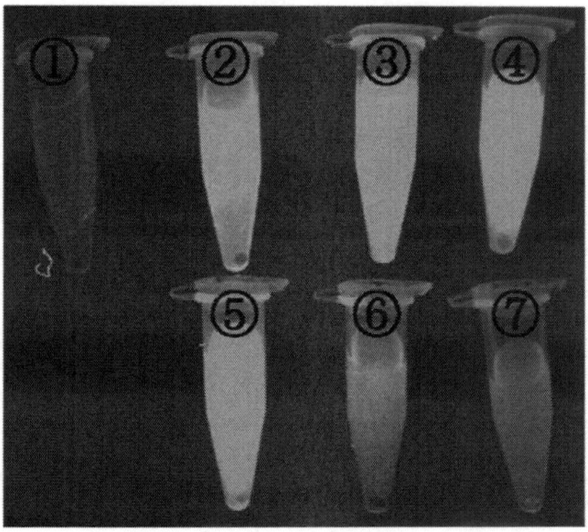

Figure 13.14. Light emission (lime green or blue fluorescence) from suspension of products generated by reductive combustion processing of collected cells in various enrichments. Cells were collected by centrifugation and then washed twice with distilled water. The collected cell paste was reductively burned at 300°C for six hours with ventilating nitrogen gas in a muffle furnace (Yamato FO 100). After burning the products were suspended with distilled water. 1: an enrichment from sediment of a pond in Imari, Saga, 2: a microbial consortium from Kinrin Lake sediment in Yufuin, Ooita, 3: SM-ER consortium (from sediment of an oil producing pond in Singleton memorial park in Kurokawa, Niigata, 4: an enrichment from a water way near Kameno river in Wakayama, 5: an enrichment from sediment of Arida river in Wakayama, 6: an enrichment from marine sediment of Osaka bay, Osaka, 7: control (*E. coli* JM109). UV at 365 nm was illuminated by a transilluminator (LX-200, TAITEC). See also Color Insert.

will be powerful bioresources for bioremediation, biomanufacturing, and metal biotechnology.

Acknowledgments

T.S. gratefully thanks Dr. Mitsufumi Matsumoto (senior director of Biotechnology Laboratory, Wakamatsu Research Institute, Electric Power Development Co. Ltd.) for corporation and scientific advices on this research. T.S. also deeply acknowledges graduates and students of SAKAGUCHI laboratory who participated and contributed in this study, Mr. Yuuki Mizoguchi, Mr. Shogo Hamada, Mr. Teruki Ueno, Mr. Takashi Imanaga, Mr. Masataka Onaka, and Mr. Ryosuke Shimizu. We also thank Prof. Eiichi Tamiya (Osaka University) and Dr. Kouichi Higashimine for cooperation in HRTEM, EDX, and electron diffraction analysis in our research. A part of this work was conducted in Kyoto-Advanced Nanotechnology Network (H20-013), supported by "Nanotechnology Network" of the Ministry of Education, Culture, Sports, Science and Technology (MEXT), Japan.

References

1. Lapage, S. P. and Bascomb, S., *J. Appl. Bacteriol.* **31**, 568–580 (1968).
2. McCready, R. G. L., Campbell, J. N., and Payne, J. I., *Can. J. Micobiol.* **12**, 703–714 (1966).
3. Tucker, F. L., Walper, J. F., Appleman, M. D., and Donohue, J., *J. Bactriol.* **83**, 1313–1314 (1962).
4. Woolfolk, C. A. and Whiteley, H. R., *J. Bactriol.* **84**, 647–658 (1962).
5. Garbisu, C., Gonzalez, S., Yang, W. H., Yee, B. C., Carson, D. L., Yee, A., Smith, N. R., Otero, R., Buchanan, B. B., and Leighton, T., *BioFactors* **5**, 29–37 (1995).
6. Lorite, L., Gould, W. D., Rajan, S., McCready, R. G. L., and Cheng, K., *Appl. Environ. Microbiol.* **58**, 4042–4044 (1992).
7. Narasingarao, P. and Häggblom, M. M., *Appl. Environ. Microbiol.* **73**, 3519–3527 (2007).
8. Moore, M. D. and Kaplan, S., *J. Bacteriol.* **174**, 1505–1514 (1992).

9. Moore, M. D. and Kaplan, S., *ASM News* **60**, 17–23 (1994).
10. Yamada, A., Miyagishima, M., Inoue, K., and Matsunaga, T., *Appl. Microbiol. Biotechnol.* **48**, 367–372 (1997).
11. Yamada, A., Miyagshima, M., and Matsunaga, T., *Mar. Biotechnol.* **5**, 46–49 (1997).
12. Oremland, R. S., Hollibaugh, J. T., Maest, A. S., Presser, T. S., Miller, L. G., and Culberson, C. W., *Appl. Environ. Microbiol.* **55**, 2333–2343 (1989).
13. Steinberg, N. A. and Oremland, R. S., *Appl. Environ. Microbiol.* **56**, 3550–3557 (1990).
14. Taylor, D. E., *Trend Microbiol.* **7**, 111–115 (1999).
15. Avazéri, C., Turner, R. J., Pommier, J., Weiner, J. H., Giordano, G., and Verméglio, A., *Microbiology* **143**, 1181–1189 (1997).
16. Basaglia, M., Toffanin, A., Baldan, E., Bottegal, M., Shapleigh, J. P., and Casalla, S., *FEMS Microbiol.* **269**, 124–130 (2007).
17. Bébien, M., Kirsch, J., Méjean, V., and Verméglio, A., *Microbiology* **148**, 3865–3872 (2002).
18. Chiong, M., González, E., Barra, R., and Vásquez, C., *J. Bactriol.* **170**, 3269–3273 (1988).
19. Hunter, W., *Curr. Microbiol.* **54**, 376–381 (2007).
20. Pierru, B., Grosse, S., Pignol, D., and Sabaty, M., *Appl. Environ. Microbiol.* **72,** 3147–3153 (2006).
21. Ridley, H., Watts, C. A., Richardson, D. J., and Butler, C. S., *Appl. Environ. Microbiol.* **72**, 5173–5180 (2006).
22. Schröder, I., Rech, S., Krafft, T., and Macy, J. M., *J. Biol. Chem.* **272**, 23765–23768 (1997).
23. Klonowska, A., Heulin, T., and Vermeglio, A., *Appl. Environ. Microbiol.* **71**, 5607–5609 (2005).
24. Soudi, M. R., Ghazvini, P. T. M., Khajeh, K., and Gharavi, S., *J. Hazard. Mater.* **165**, 71–77 (2009).
25. Cantafio, A. W., Hagen, K. D., Lewis, G. E., Bledsoe, T. L., Numan, K. M., and Macy, J. M., *Appl. Environ. Microbiol.* **62**, 3298–3303 (1996).
26. Kashiwa, M., Ike, M., Mihara, H., Esaki, N., and Fujita, M., *BioFactors* **14**, 261–265 (2001).
27. Lovley, D. R., *Annu. Rev. Microbiol.* **47**, 263–290 (1993).
28. Dameron, C. T., Reese, R. N., Mehra, R. K., Kortan, A. R., Carroll, P. J., Steigerwald, M. L., Brus, L. E., and Winge, D. R., *Nature* **338**, 596–597 (1989).

29. Williams, P., Keshavarz-Moore, E., and Dunnill, P., *J. Biotechnol.* **48**, 259–267 (1996).
30. Sharma, P. K., Balkwill, D. L., Frenkel, A., and Vairavamurthy, M. A., *Appl. Environ. Microbiol.* **66,** 3083–3087 (2000).
31. Kowshik, M., Deshmukn, N., Vogel, W., Urban, J., Kulkarni, S. K., and Paknikar, K. M., *Biotechnol. Bioeng.* **78**, 583–588 (2002).
32. Pearce, C. I., Coker, V. S., Charnock, J. M., Pattrick, R. A. D., Mosselmans, J. F. W., Law, N., Beveridge, T. J., and Lloyd, J. R., *Nanotechnology* **19**, 1–13 (2008).
33. Sakaguchi, T., Kato, M., Kuriyama, N., Niiyama, N., Hamada, S., Morita, Y., and Tamiya, E., *Curr. Microbiol.* **59**, 88–94 (2009).
34. Sakaguchi, T., Kuriyama, N., Niiyama, H., Morita, Y., and Tamiya, E., *Biomineralization (BIOM2001): Formation, Diversity, Evolution and Application, Proceedings of the 8th International Symposium on Biominelarization*, eds. Kobayashi, Y. and Ozawa, H. (Tokai University Press, Kanagawa) pp. 259–262 (2003).
35. Matsumoto, K., *Heavy Oil Spilled From Russian Tanker "Nakhodka" in 1997: Towards Eco-Responsibility, Earth Sense*, eds. Tasaki, K. (21st century COE Kanazawa University, long- and short-term dynamics of pan-Japan sea area; environmental monitoring and prediction, Kanazawa) pp. 203–213 (2003).
36. Osuji, L. C. and Onojake, C. M., *Chem. Biodivers.* **1**, 1708–1715 (2004).
37. Turner, R. J., Werner, J. H., and Taylor, D. E., *Anal. Biochem.* **204**, 292–295 (1992).

Chapter 14

BIOMACHINING–MICROMACHINING OF METALS BY BACTERIA

Yoshiyuki Uno and Akira Okada

Department of Mechanical Engineering, Okayama University, 3-1-1, Tsushimanaka, Kitaku, Okayama, Japan

uno@mech.okayama-u.ac.jp; okada@mech.okayama-u.ac.jp

14.1 Introduction

The various material processing techniques presently in use can be broadly classified into two categories according to the energy used in the technique, which are physical processing and chemical processing. Natural science system is generally classified into three categories — physical, chemical, and biological. However, no biological technique has ever been developed for material processing. The recent advances in biotechnology have led to the widespread application of its techniques in many fields. Thus, it appears desirable to investigate the possibility of biological material processing techniques. Many kinds of microorganisms exit, some of whose actions can be considered as a kind of material processing. If it is possible for microorganisms to machine the desired part to the required depth, they can be useful as tools for micromachining. From the

Handbook of Metal Biotechnology: Applications for Environmental Conservation and Sustainability
Edited by Michihiko Ike, Mitsuo Yamashita, and Satoshi Soda

www.panstanford.com

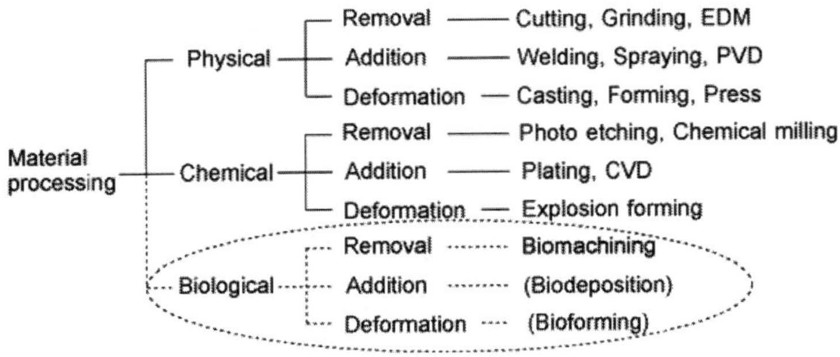

Figure 14.1. Classification of material processing.

above-mentioned viewpoint, the possibility of biological machining of metals by a kind of lithotroph, *Acidithiobacillus ferrooxidans*, which lives in a specific environment, was investigated.

14.2 Definition of Biomachining

As shown in Fig. 14.1, material processing is classified into physical, chemical, or biological processing, according to the energy used in the process and the volume change. Then each category can be classified further into removal, addition, or deformation processing. A new material processing technique using biological energy is introduced here as biomachining, biodeposition, and bioforming corresponding to three kinds of conventional material processing. Biomachining is defined as biological removal processing.

14.3 *A. Ferrooxidans*

Organisms are classified into many groups by the energy source necessary for viability and the carbon source used for nutrition as shown in Fig. 14.2.[1] Organisms which obtain carbon from organic matter are called organotrophs, and include animals, and many kinds of bacteria. On the other hand, organisms which directly utilize carbon dioxide in air as a carbon source are called lithotrophs. Plants, chlorella, and cyanophyceae are classified into

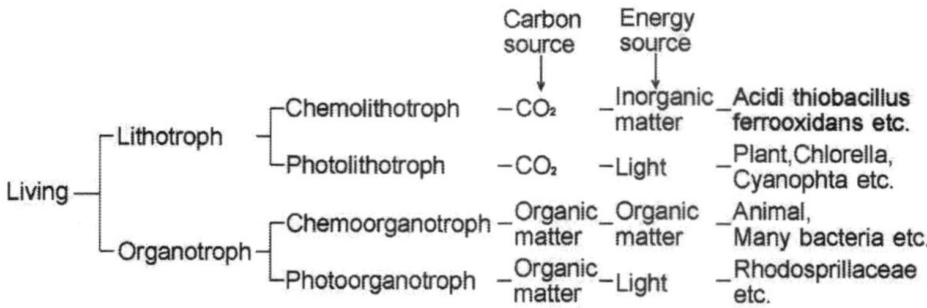

Figure 14.2. Classification of organisms from viewpoint of carbon and energy sources.

this category. From the viewpoint of energy source, organotrophs are divided into chemoorganotrophs which utilize organic matter and photoorganotrophs which use light. Similarly, lithotrophs are divided into chemolithotrophs and photolithotrophs. Among the chemolithotrophs there are some bacteria that eat metal. *A. ferrooxidans* and *Acidithiobacillus thiooxidans* are included in this category. These bacteria utilize energy generated by oxidation of iron or sulfur to fix carbon dioxide in the air. For example, *A. ferrooxidans* can gain energy by oxidation of bivalent ion as follows.

$$4Fe^{2+} + O_2 + 4H^+ \rightarrow 4Fe^{3+} + 2H_2O \tag{14.1}$$

In this reaction, about 8 kcal per mol is generated. A normal removal process using the *A. ferrooxidans* as a tool was investigated.

Figure 14.3 shows scanning electron microscope (SEM) image of *A. ferrooxidans* (ATCC13598). This is a kind of short bacillus, about

Figure 14.3. *Acidithiobacillus ferrooxidans* (ATCC13598).

0.5 μm in diameter and 1 μm in length. If it is possible to apply the bacterium to material processing it will be very advantageous as a tool for micromachining because of its small removal or deposition volume. Moreover, since the metabolic function of the bacterium is utilized, no physical or chemical energy needs to be concentrated at the machining point, so neither a damaged layer nor a heat affected zone is generated.

14.4 Culture of *A. Ferrooxidans*

There are seven known types of *A. ferrooxidans*, which live in such specific environments as waste solution from ore, registered by the American Type Culture Collection.[2] Three of them, ATCC13598, ATCC13661, and ATCC33020, were used in this investigation. These bacteria were collected from different places, but it has not been determined whether they differ from each other genetically.

For discussion on material processing by bacteria, it is necessary to obtain a pure culture of the bacteria. *A. ferrooxidans* utilizes energy generated in the oxidation of bivalent iron to trivalent iron. However, bivalent iron is oxidized spontaneously to trivalent iron at pH values smaller than 3.0. Therefore, *A. ferrooxidans* survives only in a strong acid solution with pH less than 3.0. First, an appropriate volume of 9 K medium was sterilized in an autoclave at 132°C for 12 min. Next, 3 vol.% of undiluted culture fluid was inoculated into 9 K medium. Then this medium was shaken in an incubator at 28°C, at a shaking rate of 160 cpm and with shaking width of 25 mm for both X and Y directions. The growth rate of *A. ferrooxidans* is low, since it takes about a week for the density of bacteria to reach a constant value. The number of bacteria was counted using a haemacytometer under a microscope.

Figure 14.4 shows the variations of the number of bacteria with the cultivation time. The number of bacteria ATCC13598 increases exponentially with cultivation time and reaches 10^8/mL at about 100 hours. Marked differences in growth among the three types of bacteria cultured were not confirmed. The fluid cultured for a week, in which the bacteria had reached stationary phase, was utilized in the metal removal experiments.

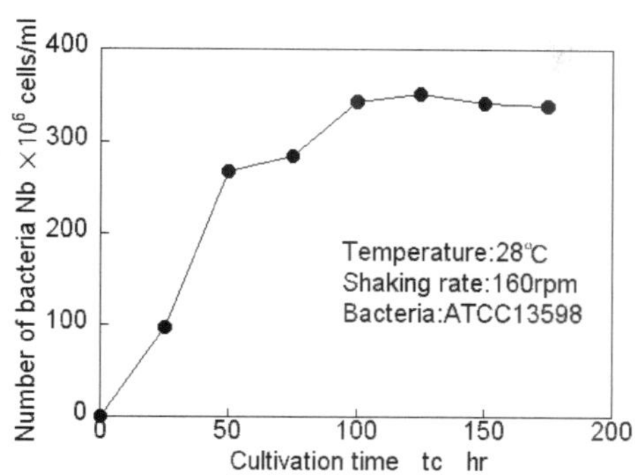

Figure 14.4. Variation of number of bacteria with cultivation time.

14.5 Metal Removal Experiment

The metal removal experiments were conducted using the apparatus shown in Fig. 14.5. The mask pattern used to form the grooves was previously prepared on the workpiece by the photolithography process. Pure iron and pure copper were adopted as the workpiece, because these bacteria have been employed to leach iron, copper, and uranium from ore.[3] The machined groove contour was measured by a surface profilemeter. Figure 14.6 shows a SEM micrograph and an example of the measurement of machined grooves. As shown in these figures, the grooves were machined with an approximately uniform depth.

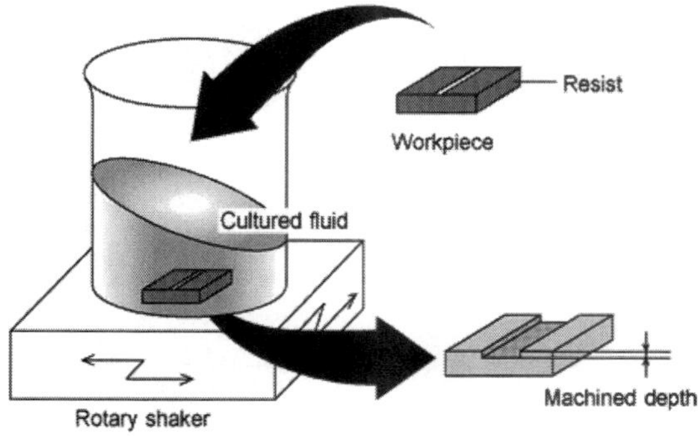

Figure 14.5. Experimental method of biomachining.

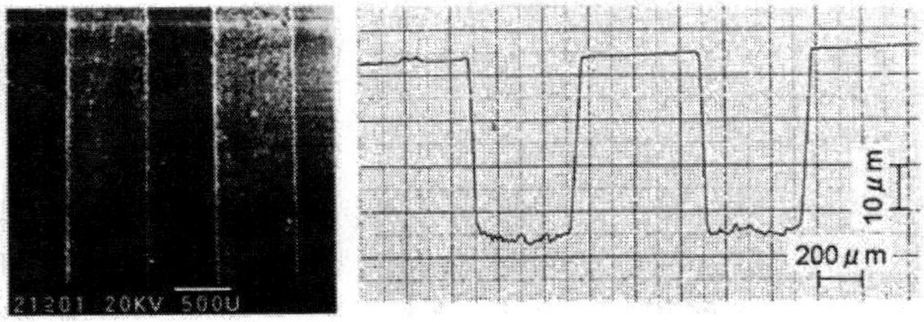

Figure 14.6. SEM micrographs and profile of machined groove.

The variation of groove depth with the machining time for pure iron is shown in Fig. 14.7. The culture fluids containing the bacteria were shaken at 28°C, 160 cpm in this experiment. The solid circles in the figure show the removal amount in 9 K medium without bacteria, which was negligible even after 10 hours. This indicates that the amount of material removal by chemical etching was minute even in a strong acid solution of pH 2.5. On the other hand, the groove depth increased almost linearly with the machining time using the culture fluid with the bacteria. This suggests that *A. ferrooxidans* can be used for machining iron and the removal amount can be controlled by the machining time. The mean removal rate for pure iron was about 14 μm per hour. As shown in the figure, there were no marked

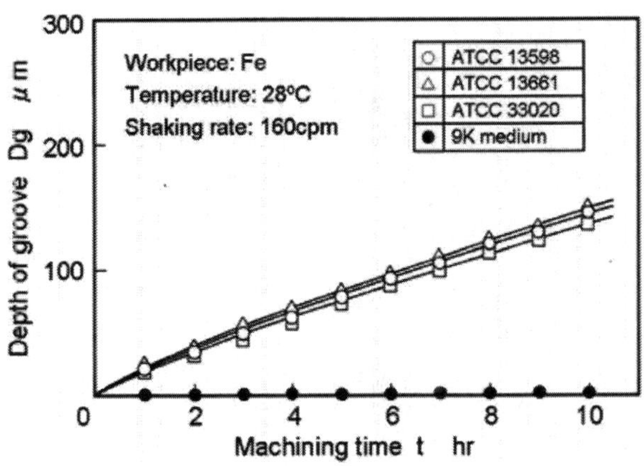

Figure 14.7. Variation of groove depth with machining time for pure iron.

differences among the three types of bacteria employed here. For pure copper, similar machining ability was confirmed and the mean removal rate for pure copper was about 20 μm per hour, which is higher than that for pure iron.

The effect of the shaking rate on the mean removal rates using bacteria ATCC13598 was also investigated. As a result, the mean removal rate reached a constant value at 160 cpm. This indicates that it is necessary for bacteria to make contact several times with the workpiece surface due to the shaking motion to reach a constant removal rate. The mean removal rate of pure copper was higher than that of pure iron for any shaking rate. Furthermore, the effects of the machining temperature on the mean removal rates were also investigated. The mean removal rate for pure copper reached its maximum around 30°C, while that for iron at around 40°C. These bacteria are optimally active around 30°C, which agrees with the result in the case of pure cupper. However, in the case of pure iron, another factor should be considered: chemical reactions which increase the removal rate with increase in the machining temperature may have occurred in addition to the bacterial activity.

14.6 Material Removal Mechanism in Biomachining

Some mineralogists have investigated the possibility of using *A. ferrooxidans* for bacteria leaching of low-grade ore,[4] particularly copper and uranium ore.[3] Two bacteria leaching mechanisms have been proposed.[5] One is the direct leaching mechanism, in which the bacteria stick to the surface of ore and oxidize the sulfur component to generate sulfuric acid. Another is the indirect leaching mechanism, in which chemical oxidation of iron allows the metal component in the ore to leach as metal sulfuric salt. For example, the leaching mechanism for uranium ore is presented as follows:

$$\begin{aligned} CaU(PO_4)_2 + Fe_2(SO_4)_3 + H_2SO_4 + 2H_2O \rightarrow UO_2SO_4 + 2FeSO_4 \\ + 2H_3PO_4 + CaSO_4 \end{aligned} \tag{14.2}$$

In this mechanism *A. ferrooxidans* has the effect of oxidizing Fe^{2+} to Fe^{3+} as shown in Eq. 14.1. It is very difficult to identify the actual leaching mechanism in practical leaching processes; both of them appear to operate simultaneously.[5] The leaching mechanism of pure metal has never been investigated before. Therefore, the mechanism of metal removal has not been clarified. However, simply considering the components of the workpiece, it is reasonably assumed that the direct leaching mechanism predominates for pure iron, while the indirect leaching mechanism predominates for pure copper.

14.7 Electric Field Assisted Biomachining

The effect of electric fields on biomachining has also been investigated since oxidation or reduction is based on the transfer of electrons. Figure 14.8 shows the variation of groove depth with the machining time when an electric field, DC 0.5 V, was applied between two workpiece of pure copper.[6] The current was maintained at 0.01 A to suppress electrochemical machining. The dotted line in the figure represents the groove depth in biomachining without the electric field, and "*b*" shows the groove depth after 10 hours. When the electric field was applied between two workpiece (shown as "*c*" for 10 hours in the figure) was about twice that in normal biomachining while that on the cathodic workpiece became extremely

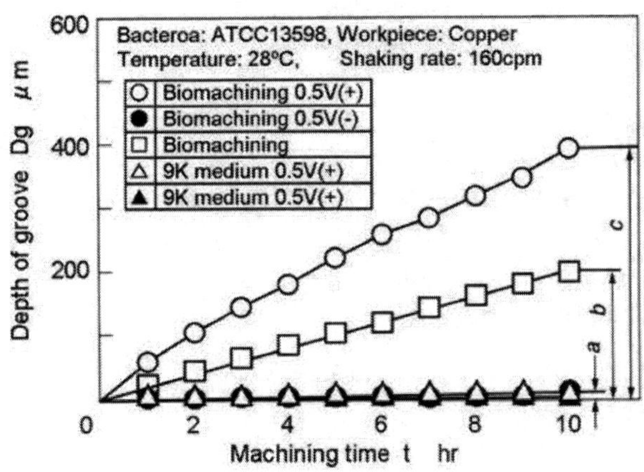

Figure 14.8. Variation of groove depth with machining in electric field assisted biomachining.

small. On the other hand, when the same electric field was applied in 9 K medium without bacteria, the anodic workpiece was slightly electrochemically machined, as shown by "*a*" in the figure. If a simple addition rule can be applied to these results, the removal amount after machining for 10 hours under the electric field should be "$a + b$." However, the removal amount c was much larger than $a + b$. This result indicates that the electric field can produce a kind of multiplication effect between biomachining and electrochemical machining. We therefore call this method "electric field assisted biomachining." A similar effect can be observed for pure iron.

14.8 Jet Biomachining and Stirring Biomachining

In the conventional biomachining methods shown above, the material removal rate is extremely low, which is very effective to precisely control the machining shape and is well applicable to micromachining. However, it is ideal to be more efficient even if the material removal is micro scale order. Therefore, some techniques were examined for achieving an efficient material removal. One of them is jet biomachining, in which cultured fluid is jetted out from nozzle onto the workpiece.[7] Figure 14.9 shows the variations of

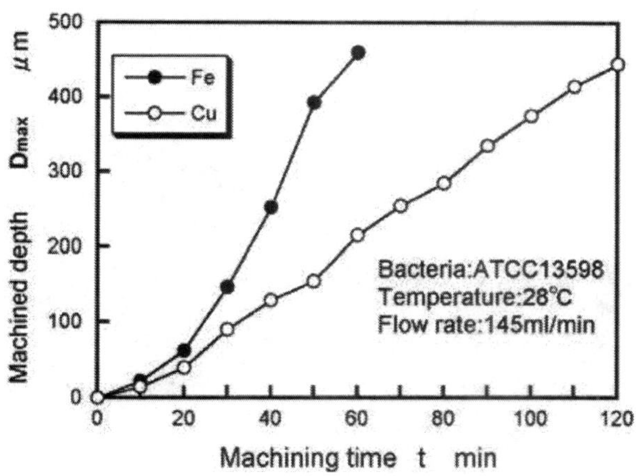

Figure 14.9. Variation of groove depth with machining in jet biomachining.

removal rate with jet flow rate. The removal rate increases with an increase of the flow rate, since the opportunity for bacteria to act on workpiece increases. The removal rate is about 450 μm per hour for pure iron and it is approximately 30 times higher than that by normal biomachining. Also when the cultured fluid in a fixed beaker is stirred with very high speed by using stirrer-stirring biomachining, very high material removal rate can also be achieved.

14.9 Culture Agar Biomachining

To obtain required shapes, the masking of no machined parts with resist film was necessary in the above-mentioned conventional method, which makes the process inefficient. A new biomachining technique using agar culture medium is proposed here for readily controlling the machining shape.[8] In this method, a cultured medium including bacteria is gelatinized by mixing agar. The hardness is varied with the amount of mixed agar, and the gelatinized culture medium with optimum hardness is very easy to make a shape by cutting it with a knife. Figure 14.10 shows one example of machined cavity by culture agar biomachining. A star-shaped gelatinized culture medium was put on the workpiece and left for 10 hours. The star shape of gelatinized medium is well transferred and good sharpness can be obtained although the machined depth is as shallow as a few microns.

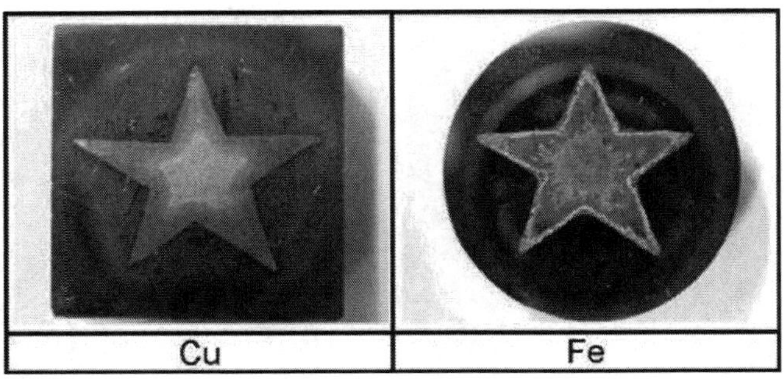

Figure 14.10. Culture agar biomachining of star shape on copper and iron.

14.10 Summary

A new metal removal process called "Biomachining" using a lithotrophic bacterium, *A. ferrooxidans*, was proposed. Experimental investigations showed that *A. ferrooxidans* in culture fluid had high ability to machine iron and copper, and the biomachining is well applicable to micromachining, since removed depth was very small but approximately proportional to the machining time. The removal rate was controlled with the shaking rate of culture fluid and its temperature. It was also clarified that removal rate at the anodic workpiece in electric field assisted biomachining, i.e. biomachining under an electric field, became much higher than that in normal biomachining, while the removal amount at the cathodic workpiece was minute. Furthermore, the removal rate could be improved by jet biomachining and stirring biomachining. In the culture agar biomachining technique, the machining shape is readily controlled by the shape of gelatinized culture medium.

Biomachining is the newborn machining process and which direction it makes progress to is unpredictable. In other words, enormous potential may be hidden in this process. Considering the recent development of genetic engineering, we strongly hope that it will be possible to incorporate a gene to play a certain roll into each bacterium in the future. In such case, we believe that many bacteria act as an autonomous-distributed micro robot in many fields.

Acknowledgments

This research was financially supported by Grants-in-Aid of the Science Research Fund and by Special Funds for Education and Research (Research Promotion) from the Ministry of Education, Culture, Sports, Science and Technology, Japan. The authors would like to thank Prof. T. Sugio, Okayama University, for supplying bacteria as well as useful information on the culture of bacteria.

References

1. Imai, K., *Lithotrophic Bacteria* (Kagaku Dojin) p. 4 (1984).
2. *American Type Culture Collection, Catalog of Bacteria and Bacteriophages*, 17th Ed., p. 250 (1989).
3. Yamanaka, T., *Energy Metabolism of Microorganisms, Gakkei Shuppan Center*, p. 148 (1986).
4. Sugio, T. and Tano, T., *Heredity* **42**(8), 28 (1988).
5. Takahashi, T. and Kai, T., *Bioindustry* **5**(11), 789 (1988).
6. Uno, Y. Kaneeda, T., and Yokomizo, S., *JSME Int. J.* **39**(4), 837 (1996).
7. Okada, A., Uno, Y., Hisano, T., and Kaneeda, T., *J. JSPE* **73**(1), 1585 (2002).
8. Okada, A., Uno, Y., Nakanishi, T., and Kaneeda, T., *J. Jpn. Soc. Adv. Prod. Tech.* **23**(1), 13 (2005).

Chapter 15

BIOSENSORS FOR TOXIC HEAVY METALS

Kazumasa Hirata[1] **and Hitoshi Miyasaka**[2]

[1] *Graduate School of Pharmaceutical Sciences, Osaka University, 1-6 Yamada-oka, Suita, Osaka 565-0871, Japan*

[2] *The Kansai Electric Power Co., Environmental Research Center, Keihanna-Plaza, 1-7 Seikacho, Sourakugun, Kyoto 619-0237, Japan*

hirata@phs.osaka-u.ac.jp; miyasaka.hitoshi@a4.kepco.co.jp

15.1 Introduction

Recently, conventional techniques for the analysis of heavy metals have undergone rapid development and have been applied to the risk management of foods and the environment, as they have for other environmental pollutants. For example, graphite furnace atomic absorption spectrometry (GF-AAS), inductively coupled plasma mass spectrometry (ICP-MS), inductively coupled plasma atomic emission spectrometry, and X-ray fluorescence spectroscopy are extremely precise analytical techniques and enable the detection of heavy metals with high sensitivity and selectivity.[2,3] However, these methods require expensive instruments with high running costs, time-consuming processes, and expert techniques. Therefore, they are disadvantageous for primary comprehensive monitoring and the on-site analysis of heavy metals in various food and environmental samples. On the other hand, biosensors are ideal

Handbook of Metal Biotechnology: Applications for Environmental Conservation and Sustainability
Edited by Michihiko Ike, Mitsuo Yamashita, and Satoshi Soda

www.panstanford.com

tools to fulfill these requirements, because they have the advantages of portability, low cost, ease of use, and rapid responses in real time, even if their sensitivity and selectivity are lower than those of the above-mentioned conventional methods.[4,5]

As is well known, biosensors are analytical devices consisting of biological materials in conjunction with a compatible transducer. Biomaterials are quantitatively responsible for the specific recognition of target analytes. Because living organisms have adapted to heavy metal poisoning during phylogenic evolution, they respond to heavy metals through various biochemical and physiological processes, thereby acquiring tolerance against them. These systems are strictly regulated at the transcriptional, translational, and post-translational levels. Therefore, the mechanisms underlying the response and adaptation to heavy metals enable us to develop useful biosensors. The transducers convert signals obtained by the interaction of the biomaterials with the analytes to various outputs, which include amplified physicochemical signals such as amperometric, potentiometric, optical, and thermal signals.

Biosensors have already been widely applied as useful monitoring systems for specific substances in the medicinal fields and food industries.[5] The commercial market of biosensors has grown in the past decade, and many attempts have been made recently to use them for environmental monitoring of various pollutants including heavy metals.[6,7] Although portability, low cost, ease of use, and rapid response in real time are advantageous properties of biosensors as mentioned above, biomaterials that produce sufficiently high sensitivity and selectivity are required as a matter of course in biosensors. Furthermore, appropriate reproducibility and stability are also required for practical use. In general, several different biomaterials, such as whole organisms, whole cells, organelles, tissues, cell membranes, and various molecules, have been used in biosensors to achieve highly sensitive and selective recognition of analytes including environmental pollutants. In the case of heavy metal analysis, however, almost all biosensors investigated so far fall into the categories of whole-cell-based or molecular-based biomaterials.[8] As shown in Table 15.1, whole-cell-based biosensors are classified as either natural or genetically engineered. On the other hand, various molecules, classified as enzymes, antibodies,

Table 15.1. Biosensors for detection of toxic heavy metals.

Classification	References
Whole-cell-based biosensor	
Non-genetically engineered microorganisms	9, 10
Genetically engineered microorganisms	11–21
Enzyme-based biosensor	
Inhibition-based enzyme	23–29
Activation-based enzyme	30, 31, our research
Antibody-based biosensor	37–41
Nucleic acid-based biosensor	
Aptamer, oligonucleotides	43–46
DNAzyme	43, 47–49
Other molecule-based biosensor	17, 50, 51

DNA, and other molecules, have been applied as biomaterials in biosensors. In this article, the biosensors that have been reported mainly in the past two decades are summarized. In addition, three simple and visible biosensors developed by our groups are introduced in detail.

15.2 Whole-Cell-Based Biosensor

Living organisms have acquired mechanisms for response and tolerance against toxic heavy metals during the course of evolution. These mechanisms can be exploited for the development of whole-cell biosensors for heavy metals. In general, both naturally occurring cells and genetically engineered cells have been employed for heavy metal detection, similar to other environmental pollutants. Although there have been several reports concerning whole-cell biosensors based on naturally occurring cells,[9,10] genetically engineered cells are dominant in heavy metal analysis. Genetically engineered whole-cell biosensors are constructed through the genetic fusion of a promoter and a reporter gene. The recombinant gene is located on either a plasmid or the chromosome. The promoter responds to heavy metals invading the cells and induces the expression of the reporter gene. Expression occurs in proportion to the concentration of heavy metals and can be quantitatively detected by optical or electrochemical reactions. The sensitivity and selectivity of biosensors are dependent on the choice of these two constituents and their combinations.

In general, the microbial gene promoters employed for biosensors recognize heavy metals with high sensitivity. Well-known promoters for toxic heavy metals include *cad* for Cd,[11] *mer* for Hg,[12] and *ars* for As.[13] However, the selectivity of the promoters is often not very high. For example, promoters for Cd detection are usually also responsive to other heavy metals such as Cu, Hg, and Zn.[14] As described in Ref. 6, novel promoters that response to certain heavy metals can be identified from appropriate microorganisms and utilized in heavy metal biosensors by using transcriptomic or proteomic technologies.

Various different reporter genes have been used in heavy metal biosensors in combination with the above promoters. Similar to biosensors used for other analytes, genes encoding enzymes such as β-galactosidase (*lacZ*),[14,15] bacterial luciferase (*lux*), and firefly luciferase (*luc*)[13] have been applied as reporter genes for the monitoring of heavy metals. The intensity of specific colors or fluorescence produced by these enzymes is proportional to the level of gene expression and can be significantly amplified by the enzyme catalytic reaction. However, these enzyme reactions require sufficient amounts of substrate. Usually, therefore, an adequate amount of substrate must be added to the cells. Genes encoding fluorescent proteins such as GFP[16] and luxCDABE reporter system[17] are also available as reporter genes. These optical signals, which are in proportion to the level of gene expression, can be directly detected by visible spectrophotometer or luminometer, but significant signal amplification is unlikely.

In general, the sensitivities of whole-cell heavy metal biosensors are sufficiently comparable to those of conventional detection methods such as GF-AAS and ICP-MS. For example, nanomolar and femtomolar levels of the toxic heavy metals Cd, Cu, Hg, and Pb can be detected.

Maeda *et al.* have developed whole-cell As biosensors using carotenoid biosynthesis genes. The reporter gene employed was *crtA*, which encodes an enzyme catalyzing demethylspheroidene or sheroidene to demethylspheroidenone or sheroidenone,

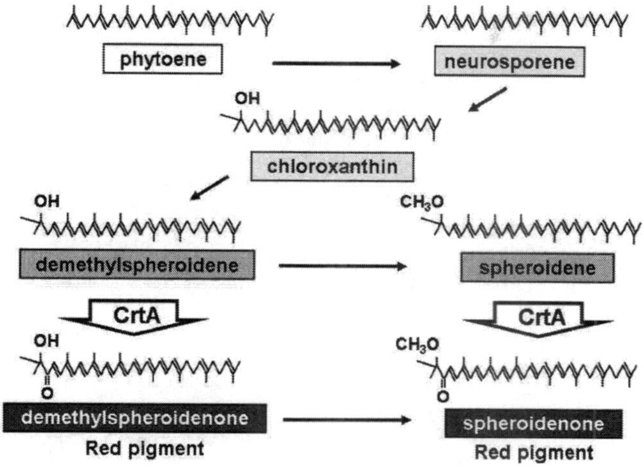

Figure 15.1. Carotenoid synthesis in spheroidene pathway. See also Color Insert.

respectively, in the carotenoid biosynthetic pathway of a photosynthetic bacterium, *Rhodovulum sulfidophilum* (Fig. 15.1). An As biosensor plasmid, containing the operator/promoter region of the *ars* operon, the *arsR* gene from *Escherichia coli*, and this gene, was constructed in *E. coli*.[18,19] The substrate color, yellow, was changed to the product color, red, which could be easily detected by naked eye with a sensitivity of 5 μg/L As. Recently, Maeda *et al.* developed a more sensitive biosensor by using the *crtI* gene from *Rhodopseudomonas palustris* instead of *crtA*.[20,21] In this case, CrtI catalyzed the conversion of phytoene to lycopene. Lycopene was further converted to other carotenoids with red color. Consequently, the green-yellow color of bacteria was changed to red, which could be distinguished more clearly than the previous system (Fig. 15.2). These biosensors are promising for applications to on-site primary comprehensive monitoring of As, because they achieve high sensitivity without using any instruments or the addition of any reagents.

The advantages of whole-cell bacterial biosensors are inexpensive, quick, and easy to measure, sensitive, less labor intensive, suitable for *in situ* analysis. On the other hand, a major disadvantage is short lifetime and lack of stability in comparison with other biosensors.

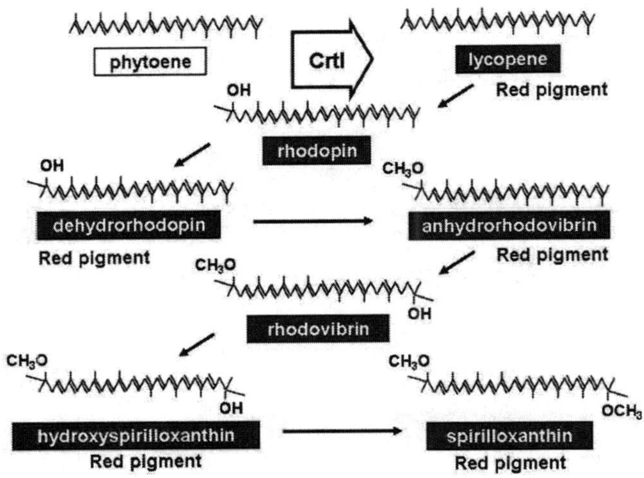

Figure 15.2. Carotenoid synthesis in spirilloxanthin pathway. See also Color Insert.

15.3 Enzyme-Based Biosensors

Not surprisingly, enzymes were among the first biomaterials incorporated into biosensors. Enzymes recognize various target analytes as substrates with high sensitivity and selectivity. The application of glucose oxidase to a glucose sensor was probably the first and most representative report of the development of an enzyme-based biosensor.[22] Enzymes can be utilized as biomaterials in biosensors on the basis of not only their recognition of analytes as substrates, but also the induction or inhibition of their activities by analytes. For the detection of toxic heavy metals, the latter mechanism is mainly employed, because heavy metals are rarely recognized by enzymes as substrates. On the other hand, the effects of heavy metals on enzyme activities have been well studied and have been applied to the development of biosensors. In general, a heavy metal binds to a specific site present in the enzyme and causes conformational and functional changes affecting enzyme activity. In the case of toxic heavy metals, inhibition-based biosensors are more common than activation-based ones. For example, peroxidase for detection of Hg,[23,24] invertase for Hg,[25] urease for Hg,[26] and other metals,[27] several oxidases for Hg,[28] and lactate dehydrogenase for various metals[28,29] have been used for the detection of toxic heavy metals with the sensitivities in the range of 0.1–10 μM. The main disadvantage of inhibition-based biosensors is their low selectivity. In many cases, the metal-binding site of an enzyme such as thiol does not strictly recognize the target heavy metal and also binds to other metals, although the binding intensities differ among the metals and depend on the three-dimensional structures of enzymes, resulting in a specific pattern in their binding.

In the case of activation-based biosensors, more selective detection of the target metals, as compared with the above-mentioned inhibition-based biosensors, can be expected because these enzymes require specific metals as co-factors for activation. Although other metals may also show some binding intensity for the enzyme, their affinity is commonly much lower than that of the target metal. In both inhibition- and activation-based biosensors, a signal intensity proportional to the concentration of the target metal can be expected, just as in enzyme biosensors based on the recognition of analytes as substrates, because conversion of the substrate to the product occurs in proportion to the activity of the enzyme that is inhibited or activated by the metal. However, the signal from the transducer in activation-based biosensors is thought to be more easily and sensitively detected, because it increases from no background at zero concentration in direct proportion to the level of the target metal. In the case of inhibition-based biosensors, by contrast, the signal intensity is in inverse proportion to the metal level and the high background at zero makes it difficult to detect trace levels of metals. The most representative report of an activation-based heavy metal biosensor is that for Zn detection using an alkaline phosphatase.[30,31] A covalently immobilized enzyme was employed in this biosensor, and highly selective and sensitive detection at a concentration of 0.02 ppb was achieved.[30]

The main disadvantage of activation-based biosensors is their limitation for use in the presence of essential heavy metals such as Zn. Specific enzyme activation by non-essential toxic heavy metals such as As, Cd, Hg, and Pb is rarely found, although their competitive or non-specific inhibition of enzymes has been well studied. In higher plants, when cells are exposed to toxic heavy metals, a small heavy-metal-binding peptides, called phytochelatins (PCs), are inducibly synthesized for their detoxification. Grill *et al.* confirmed that PC synthesis is catalyzed by PC synthase (PCS, EC2.3.2.15), and PCS is activated in the presence of several kinds of heavy metal.[32] Therefore, the activation of PC synthesis by heavy metals can be applied to activation-based heavy metal biosensors.

PCS catalyzes the biosynthesis of PCs from the substrate glutathione (GSH). GSH is converted to γglutamylcysteine (γEC) and glycine in the first reaction, and the γEC that is synthesized then binds to another GSH molecule to synthesize PC_2 in the second reaction. The product of the first reaction, γEC, also binds to PC_n to synthesize PC_{n+1}, and thus the units of γEC in the PC molecules are extended one by one. Because the first and second reactions both strictly require the presence of heavy metals and the first reaction is the limiting step in the PC biosynthetic pathway, the increase in the PC synthetic rate is expected to be proportional to the concentration of activator heavy metal. Oven *et al.* showed that Cd is by far the strongest activator of *Arabidopsis thaliana* PCS (AtPCS1), as compared with other heavy metals including As, Cu, Hg, and Zn, and light elements such as Ca, K, Li, Mg, and Na do not activate it.[33] Therefore, we attempted to develop a novel enzyme activation-based biosensor for detecting Cd using a recombinant protein of AtPCS1 as follows.

Quantitative detection of PCs has been performed by derivatization with free sulfhydryl (SH) labeling reagents, such as 5′-dithiobis (2-nitrobenzoic acid) and monobromobimane. However, these methods cannot distinguish other thiol compounds such as GSH from PCs, and require a subsequent separation process with reverse-phase HPLC.[34] As an alternative PC-specific detection method for Cd biosensors, we employed the novel fluorescence derivatization of PCs as the product of AtPCS1. Dipyrenylalkanes are known to form intramolecular excimers, in which one excited pyrene can form an excited-state complex with the other, ground-state pyrene in the molecule.[35] The excimer emits longer wavelength fluorescence at 450–500 nm than pyrene monomer fluorescence at 350–400 nm. Thus, a highly sensitive and selective detection method for PCs has been developed as follows. A pyrene derivative, *N*-(1-pyrenyl)maleimide (NPM), binds to thiol compounds via the free SH group. In the case of PC_2, the reactions shown in Fig. 15.3 are expected, where formation of the intramolecular excimer occurs from two molecules of NPM and one molecule of PC_2 with its two SH groups. On the other hand, excimer formation cannot occur between NPM and GSH, because GSH has only one free SH group. When 750 μM NPM and 100 μM PC_2 were reacted, strong fluorescence at 450–520 nm was observed after irradiation with excitation light at 345 nm. As expected, such emission of fluorescence was not obtained in the case of GSH, and the difference between the two reactions could be clearly distinguished, even by the naked eye (Fig. 15.4). It was confirmed that the activity of AtPCS1 increases in proportion to a Cd concentration in the range of 0.5–100 μM, and the fluorescence intensity also increases proportionally to PC_2. Furthermore, almost the only product of an *in vitro* PC synthetic reaction from GSH catalyzed by AtPCS1 was PC_2, and PC_2 gave the strongest fluorescence after reaction with NPM among the various PC molecules tested, including PC_3 and PC_4. In fact, when NPM-labeled GSH is used as the substrate of AtPCS1, a very simple assay process can be established, in which samples are directly added to a mixture of substrate and enzyme, and the resulting fluorescence can be observed by the naked eye after 30–60 minutes. This plant enzyme can catalyze the reaction in a temperature range of

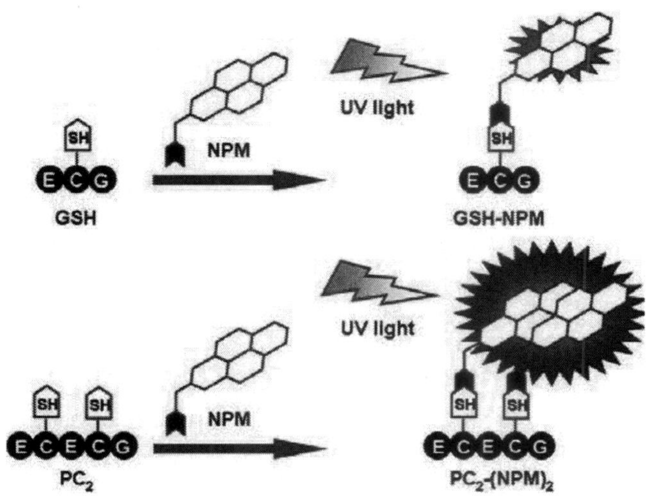

Figure 15.3. Intramolecular excimer-forming fluorescence derivatization of PC_2 after reaction with NPM.

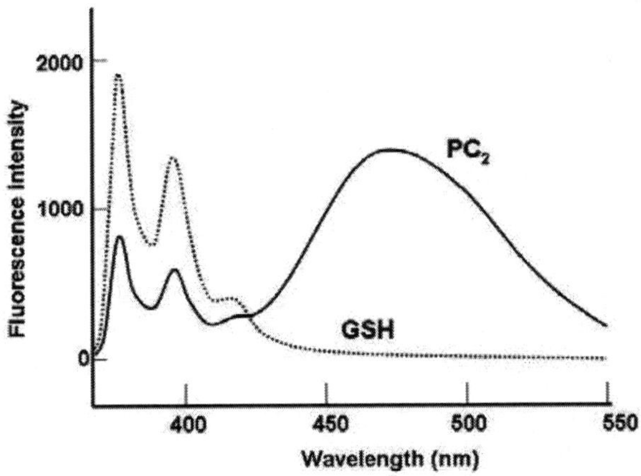

Figure 15.4. Fluorescence emission spectra (excitation 345 nm) of 100 μm PC_2 and GSH after reaction with NPM.

4–37°C. On the other hand, the disadvantages of this Cd biosensor are low sensitivity (ca. 1 μM) and low selectivity (ca. 50% cross reaction with Cu). Not only the activity of PCS but also the activation intensity of PCS by several toxic heavy metals has been found to differ considerably among the higher plants and other organisms containing this enzyme that have been so far investigated. Indeed, we have found and characterized strong Zn-induced PC synthesis in a marine green alga, *Dunaliella tertiolecta*.[36] Therefore, it will be useful to find a more sensitive, selective, or stable PCS to improve the properties of this Cd biosensor. Nevertheless, this biosensor is thought to be applicable to primary comprehensive on-site analysis of Cd in several fields. For example, Cd can be monitored in polished rice to a sufficient sensitivity of over 0.4 ppm (3.6 μM), a level that has been proposed as an international standard in foodstuffs by the Codex Alimentarius Commission of FAO and World Health Organization (WHO).

15.4 Antibody-Based Biosensors

In general, for highly sensitive and selective detection of analytes, antibody-based biosensors are more promising than biosensors utilizing other biomaterials including enzymes. Theoretically, antibodies can be generated against any organic compound including environmental pollutants. In the medical field, in particular, monoclonal antibodies are frequently used in biosensors for both monitoring medicines and detecting disease bio-markers such as hormones and enzymes.[5] In the case of toxic heavy metals, however, there have not been many reports of antibody-based biosensors, because antibodies against heavy metals alone cannot be generated. It is well known, however, that an ethylenediaminetetraacetic acid (EDTA)-chelated heavy metal can be recognized as an antigen, and this process has been employed to generate monoclonal antibodies against various toxic heavy metals.[37,38] In addition to EDTA, diethylenetriamine pentaacetic acid and cyclohexyldiethylenetriamine-pentaacetic acid have been used to make complexes with heavy metals for the generation of antibodies.[39,40] Nanomolar levels of Cd, Co, and Pb can be detected by these immunoassays.

The Kansai Electric Power Co., Central Research Institute of Electric Power Industry (CRIEP) and Sumika Chemical Analysis Service Ltd." of Japan has developed a simple biosensor for the detection of Cd named "Cadmierre." As mentioned above, a Cd-EDTA complex was used as an antigen to obtain monoclonal antibodies. Cadmierre is an immunochromatographic-type assay kit that incorporates a measuring block based on the antigen (Cd-EDTA) and antibody reaction (Fig. 15.5). The antibody is labeled with gold colloid. When a sample solution is dropped onto a sheet of nitrocellulose paper, a red band appears within 30 minutes, and the Cd concentration can be estimated by the naked eye from the density of the band. Alternatively, the density can be quantitatively digitalized with a dedicated reader. This method facilitates the detection of Cd in ppb units, and the linear range for quantitative detection is 10 ppb to 100 ppb. Cross reactivity with other heavy metals is less than 2%. Therefore, this method is promising for on-site monitoring of Cd in foodstuffs and environmental samples. It has been confirmed that removal of Mn, Mg, and Zn by pretreatment with an anion-exchange column is necessary when this method is applied to rice grain, because the concentrations of these three metals are commonly too high to neglect their cross reactivity with Cd.[41,42]

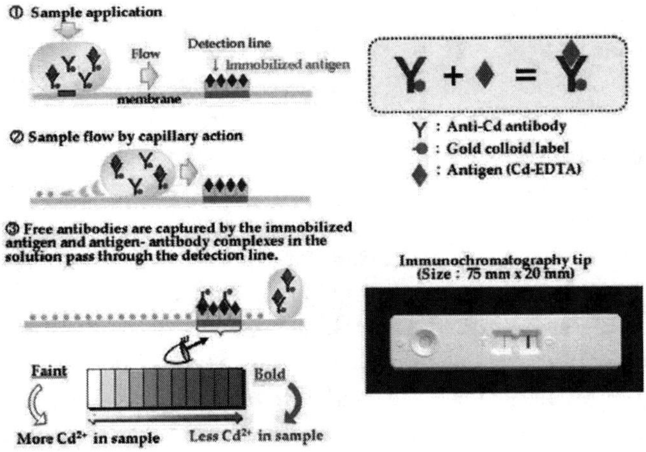

Figure 15.5. Immunochromatography for detection of cadmium. See also Color Insert.

15.5 Nucleic Acid-Based Biosensors

As well as antibodies, DNA is also a promising biomaterial for biosensors. The interaction between double-stranded DNA and low-molecular-weight environmental pollutants including heavy metals can be applied to biosensors. DNA aptamers are another type of promising material for the recognition of a wide range of analytes, because theoretically they can selectively bind to any target of choice, even to low-molecular-mass compounds that cannot generate antibodies. The toxic heavy metals Cu, Hg, Pb, and Zn can be detected by using DNA aptamers with high sensitivities.[43–45] Sensitive and selective Hg detection was also achieved by a biosensor using thiol-functionalized *poly-T* oligonucleotides.[46] Recently, a catalytically active DNA, named DNAzyme, was utilized in a heavy metal biosensor and showed highly sensitive and specific recognition of Cu, Pb, and Zn.[43,47–49]

15.6 Biosensors Based on other Molecules

Non-enzymatic metal-binding peptides and proteins, including naturally occurring and genetically engineered polypeptides, have been utilized in the development of biosensors. PCs are low-molecular-weight, heavy metal-binding peptides as mentioned above. Naturally occurring PCs, $(\gamma\text{Glu-Cys})_n$-Gly, are synthesized enzymatically, but genetically engineered PC derivatives, $\gamma\text{Glu-Cys})_n$-Gly, can be synthesized. The fusion protein of $(\alpha\text{Glu})_{20}$-Gly with maltose-binding domain protein has been prepared as a novel capacitance heavy metal biosensor. Cd, Cu, Hg, Pb, and Zn have been sensitively detected in the concentration range of 100 fM to 10 mM.[50] Metallothioneins (MTs) are ubiquitous low-molecular-weight proteins with strong binding capacity for heavy metals, including mainly Cd, Cu, and Zn. MTs have been found in many different organisms from mammals to prokaryotic microorganisms. A fusion protein of a synechoccocal MT, SmtA, with GSH-S-transferase has also been utilized in a capacitance heavy metal biosensor. In addition to Cd, Cu, and Zn, Hg was detected with high sensitivity around the femtomolar concentrations.[17] Application of MTs to biosensors has been reviewed in Ref. 51 in detail. In the case of these metal-binding peptides and proteins, however, high selectivity cannot be expected for reasons similar to those outlined for the PCS-based Cd biosensor described above.

15.7 Conclusions

This article has outlined several biosensors based on different biomaterials responsible for the specific recognition of toxic heavy metals. Although biosensors are very important tools for the risk management of foods and the environment, there have been not many biosensors available to practical use including on-site real time monitoring. Recently, for example, several biosensors enable the detection of heavy metals with high sensitivity and selectivity sufficiently comparable to or over those of conventional detection methods. More flexible and potable biosensors have been developed. Furthermore, extensive researches have being attempted to overcome the disadvantageous properties of individual biomaterials and signal transducer systems. In the near future, therefore, biosensors will be promising for effective comprehensive monitoring of toxic heavy metals and for contribution to minimize their public health impact.

References

1. Codex 2006 Codex Report of the 29th session of the Codex Alimentarius Commission. Codex Alimentarius Commission, Alinorm 06/29/41 (2006).
2. Balaram, V., *Bull. Mater. Sci.* **28**, 345–348 (2005).
3. Jackson, K. W. and Chen, G., *Anal. Chem.* **68**, 231R–256R (1996).
4. Gronow, M., *Trends Biochem. Sci.* **9**, 336–340 (1984).
5. Rich, R. L. and Myszka, D. G., *J. Mol. Recogn.* **21**, 289–312 (2008).
6. Ron, E. Z., *Curr. Opin. Biotechnol.* **18**, 252–256 (2007).
7. Wanekaya, A. K., Chen, W., and Mulchandani, A., *J. Environ. Monit.* **10**, 703–712 (2008).
8. Verma, N. and Singh, M., *Biometals* **18**, 121–129 (2005).
9. D'Souza, S. F., *Biosens. Bioelectron.* **16**, 337–353 (2001).
10. Riedel, K., Kunze, G., and Konig, A., *Adv. Biochem. Eng. Biotechnol.* **75**, 81–118 (2002).
11. Tauriainen, S., Karp, M., Chang, W., and Virta, M., *Biosens. Bioelectron* **13**, 931–938 (1998).
12. Bontidean, I., Mortari, A., Leth, S., Brown, N. L., Karlson, U., Larsen, M. M., Vangvonsveld, J., Corbisier, P., and Csoregi, E., *Environ. Pollut.* **131**, 255–262 (2004).
13. Tauriainen, S., Virta, M., Chang, W., and Karp, M., *Anal. Biochem.* **272**, 191–198 (1999).
14. Biran, I., Babai, R., Levcov, K., Rishpon, J., and Ron, E. Z., *Environ. Microbiol.* **2**, 285–290 (2000).
15. Peitzsch, N., Eberz, G., and Nies, D. H., *Appl. Environ. Microbiol.* **64**, 453–458 (1998).
16. Chakraborty, T., Babu, P. G., Alam, A., and Chaudhari, A., *Curr. Sci.* **94**, 800–805 (2008).
17. Corbisier, P., Lelie, D., Borremans, B., Provoost, A., Lorenzo, V., Brown, N. L., Lloyd, J. R., Hobman, J. L., Csöregi, E., Johansson, G., and Mattiasson, B., *Anal. Chim. Acta* **387**, 235–244 (1999).
18. Maeda, I., Yamashiro, H., Yoshioka, D., Onodera, M., Ueda, S., Kawase, M., Miyasaka, H., and Yagi, K., *Appl. Microbiol. Biotechnol.* **70**, 397–402 (2006).
19. Fujimoto, H., Wakabayashi, M., Yamashiro, H., Maeda, I., Isoda, K., Kondoh, M., Kawase, M., Miyasaka, H., and Yagi, K., *Appl. Microbiol. Biotechnol.* **73**, 332–338 (2006).

20. Yoshida, K., Yoshioka, D., Inoue, K., Takaichi, S., and Maeda, I., *Appl. Microbiol. Biotechnol.* **76**, 1043–1050 (2007).
21. Yoshida, K., Inoue, K., Takahashi, Y., Ueda, S., Isoda, K., Yagi, K., and Maeda, I., *Appl. Environ. Microbiol.* **74**, 6730–6738 (2007).
22. Updike, S. J. and Hicks, G. P., *Nature* **214**, 986–988 (1967).
23. Shekhovtsova, T. N., Muginova, S. V., and Bagirova, N. A., *Anal. Chim. Acta* **344**, 145–151 (1997).
24. Han, S., Zhu, M., Yuan, Z., and Li, X., *Biosens. Bioelectron.* **16**, 9–16 (2001).
25. Mohammadi, H., Amine, A., Ouarzane, A., and Rhazi, M. E., *Microchim. Acta* **149**, 251–257 (2005).
26. Andres, R. T. and Narayanaswamy, R., *Analyst* **120**, 1549–1554 (1995).
27. Krawczyk, T. K., Moszczyn'ska, M., and Trojanowicz, M., *Biosens. Bioelectron.* **15**, 681–691 (2000).
28. Gayet, J. C., Haouz, A., Meyer, A., and Burstein, C., *Biosens. Bioelectron.* **8**, 177–183 (1993).
29. Fennouh, S., Casimiri, V., Meyer, A. G., and Burstein, C., *Biosens. Bioelectron.* **13**, 903–909 (1998).
30. Kamtekar, S. D., Pande, R., Ayyagari, M. S., Marx, K. A., Kaplan, D. L., Kumar, J., and Tripathy, S., *Anal. Chem.* **68**, 216–220 (1996).
31. Satoh, I., *Biosens. Bioelectron.* **6**, 375–379 (1991).
32. Grill, E., Löffler, S., Winnacker, E.-L., and Zenk M. H., *Proc. Natl. Acad. Sci. USA* **86**, 6838–6842 (1989).
33. Oven, M., Page, J. E., Zenk, M. H., and Kutchan, T. M., *J. Biol. Chem.* **277**, 4747–4754 (2002).
34. Hirata, K., Tsujimoto, Y., Namba, T., Ohta, T., Hirayanagi, N., Miyasaka, H., Zenk, M. H., and Miyamoto, K., *J. Biosci. Bioeng.* **92**, 24–29 (2001).
35. Nohta, H., Satozono, H., Koiso, K., Yoshida, H., Ishida, J., and Yamaguchi M., *Anal. Chem.* **72**, 4199–4204 (2000).
36. Tsuji, N., Hirayanagi, N., Okada, M., Miyasaka, H., Hirata, K., Zenk, M. H., and Miyamoto, K., *Biochem. Biophys. Res. Comm.* **293**, 653–659 (2002).
37. Blake, D. A., Chakrabarti, P., Khosraviani, M., Hatcher, F. M., Westhoff, C. M., Goebel, P., Wylie, E., and Blake R. C. II, *J. Biol. Chem.* **271**, 27677–27685 (1996).
38. Khosraviani, M., Pavlov, A. R., Flowers, G. C., and Blake, D. A., *Environ. Sci Tech.* **32**, 137–142 (1998).

39. Blake, D. A., Jones, R. M., Blake, R. C. II, Pavlov, A. R., Darwish, I. A., and Yu, H., *Biosens. Bioelectron.* **16**, 799–809 (2001).
40. Blake, D. A., Jones, R. M., Blake, R. C. II, Pavlov, A. R., Darwish, I. A., and Yu, H., *Biosens. Bioelectron.* **16**, 799–809 (2001).
41. Abe, K., Sakurai, Y., Okuyama, A., Sasaki, K., and Tawarada, K., *J. Sci. Food Agric.* **89**, 1097–1100 (2009).
42. Sasaki, K., Yongvongsoontorn, N., Tawarada, K., Ohnishi, Y., Arakane, T., Kayama, F., Abe, K., Oguma, S., and Ohmura, N., *J. Agric. Food Chem.* **57**, 4514–4519 (2009).
43. Lui, J., Cao, Z., and Lu, Y., *Chem. Rev.* **109**, 1948–1998 (2009).
44. Rajendran, M. and Ellington, A. D., *Anal. Bioanal. Chem.* **390**, 1067–1075 (2008).
45. Kawakami, Y., Siddiki, M. S., Inoue, K., Otabayashi, H., Yoshida, K., Ueda, S., Miyasaka, H. and Maeda I., Biosens. Bioelectron., **15**, 1466–1473 (2010).
46. Cao, R.-G., Zhu, B., Li, J., and Xu, D., *Electrochem. Comm.* **11**, 1815–1818 (2009).
47. Lui, J. and Lu, Y., *J. Am. Chem. Soc.* **125**, 6642–6643 (2003).
48. Breaker, R. R., *Curr. Opin. Biotechnol.* **13**, 31–39 (2002).
49. Zuo, P., Yin, B.-C., and Ye, B.-C., *Biosens. Bioelectron.* **25**, 935–939 (2009).
50. Bontidean, I., Ahlqvist, J., Mulchandani, A., Chen, W., Bae, W., Mehra, R. K., Mortari, A., and Csoregi, E., *Biosens. Bioelectron.* **18**, 547–553 (2003).
51. Gutierrez, J. C., Amaro, F., and M.-Gonzalez, A., *BioEssays* **31**, 805–810 (2009).

Chapter 16

CELL SURFACE DESIGN FOR SELECTIVE RECOVERY OF RARE METAL IONS

Kouichi Kuroda and Mitsuyoshi Ueda
Division of Applied Life Sciences, Graduate School of Agriculture, Kyoto University, Sakyo-ku, Kyoto 606-8502, Japan
k_kuro@kais.kyoto-u.ac.jp

16.1 Introduction

In modern industrial society, the use of inorganic metal elements such as rare metals is essential. At the same time, industrial depression by unprecedented environmental pollution and depletion of resources is a growing concern. In particular, rare metals have been attracting public attention along with the development of civilized societies. They are critically important materials for human health, economy, and diplomacy between countries. A stable supply of rare metals is very critical for the development of global science and technology. Issues such as rare metal panic and alternative elemental strategies are at the top of the agenda of most governments. Therefore, innovative technologies for efficient separation and recovery of rare metals from discarded high-tech products and wastewater have to be developed. Metal adsorption using biological function

Handbook of Metal Biotechnology: Applications for Environmental Conservation and Sustainability
Edited by Michihiko Ike, Mitsuo Yamashita, and Satoshi Soda

www.panstanford.com

is expected to have advantages in individually specific adsorption and concentration recovery that is beyond the reach of conventional methods. Cell surface design is a novel strategy of molecular breeding of bioadsorbents. By cell surface engineering, metal-binding proteins and peptides were displayed on the yeast cell surface.[1–4] This system enables us not only to remove toxic heavy metals but also to adsorb and recover useful rare metals from the environment.[5] In this chapter, a novel approach by cell surface design for adsorption and recovery of rare metal ions is attempted, and the advantages and potential of cell surface design in the construction of resource-recycling bioadsorbents are described.

16.2 Social Significance of Metals

The image of metals as environmental pollutants hitherto has been accounted for most viewpoints of metals. Disposal of metals leads to environmental pollution, but collected metals could be resources. Therefore, an efficient technique of recovering metals is important. In comparison with other metals, rare metals are globally scarce, difficult to be substituted with others, and unevenly distributed in only a few regions, such as Africa, China, Russia, and South America. Rare metals with some useful features are crucial for high-tech industries, such as those producing automobiles, digital consumer electronics, and information-related devices, and significantly contribute to downsizing, weight saving, high performance, and energy saving. Therefore, rare metals are called the vitamins of such industries, and demands for such metals are increasingly growing as being indispensable for these high-tech industries to maintain and develop their international competitiveness. However, the price of rare metals is apt to fluctuate because of a weak supply system. China, which is the top exporting country, has started restricting of the export of rare metals owing to increasing domestic demand with its surging economy. Other BRICs countries (Brazil, Russia, and India) also show increasing demand with economic growth. Actually, the international price of rare metals has greatly increased in recent years. Because the amount of rare metal deposit on earth is finite, its supply is threatened by resource depletion and price increases.

Therefore, a stable supply and securement of rare metals are important issues for economic activity. On the other hand, a large amount of rare metals are accumulated in high-tech products, and rare metals in discarded products should be regarded as resources. In this respect, the stock of discarded rare metals existing in a society is called an "urban mine." As one of the solutions to this risk of unstable rare metal supply, urban mines are attracting attention. This means that there should be an effective recovery system for rare metals already existing in discarded electric devices. For the reasons stated above, the recovery and recycling of rare metals is a promising social system. Toward the realization of this system, a technology for efficient utilization of resources by recovering rare metals from nature, wastewater, and urban mines is required.

16.3 Cell Surface Adsorption for Recovery of Metal Resources

Cell surface design is a useful approach for the construction of novel bioadsorbents. Cell-surface-engineered yeast constructed by cell surface engineering has some advantages compared with conventional bioadsorbents.[1–5] The small size of microorganisms such as yeast provides a high ratio of surface area to volume and thus a large contact area interacting with metal ions in the surrounding environment.[6] Cell surface adsorption of metal ions could be performed in a short time and allows the easy recovery and recycling of adsorbed metal ions without disintegration of cells (Fig. 16.1). This is important in considering how to recover the adsorbed metal ions from cells after bioadsorption. This leads to the sustainability of natural resources in modern societies. Cell-surface-engineered yeast could be repeatedly utilized as a bioadsorbent, owing to less damage to cells during the recovery of adsorbed metal ions. In addition, the maintenance of function of metal-binding proteins on the cell surface contributes to the adsorption ability of cells, even if cells are nonviable. Thus, they are economically advantageous Cell-surface-engineered yeast constructed by cell surface display of metal-binding proteins plays a dual role of a carrier and a producer of metal-binding proteins. Consequently, the production and

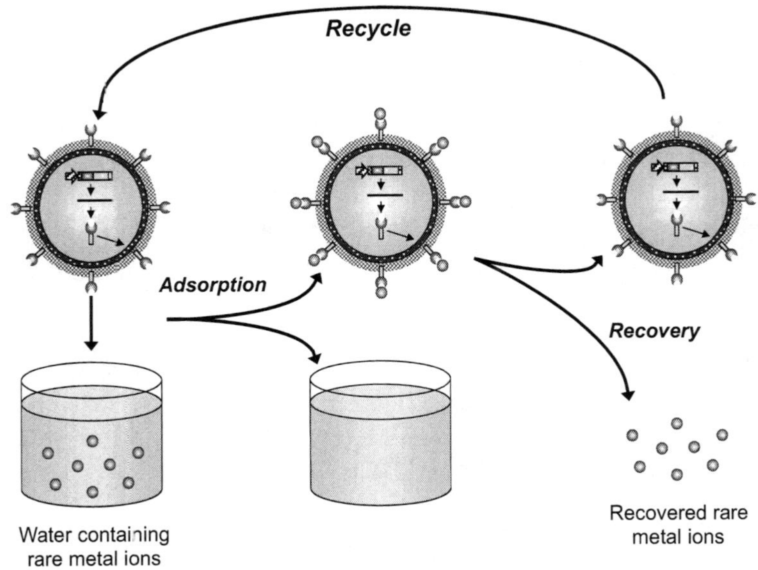

Figure 16.1. Adsorption, recovery, and recycling system of metal ions using cell-surface-engineered yeast. See also Color Insert.

conjugation of metal-binding proteins could be achieved by a simple procedure such as cell cultivation. The metal-binding ability of cell-surface-engineered yeast depends on the feature of the displayed metal-binding proteins. Therefore, the cell surface display of proteins showing selective metal binding makes it easier to endow cells with the selective adsorption of target metal ions. This is a very important point because cell surface adsorption could be applicable to the recovery of rare metals as well as the removal of toxic metal ions. By replacing the metal-binding proteins to be displayed according to the target metal, cell-surface-engineered yeast could be used for various metal resources on earth.

16.4 Molybdate Recovery by Cell Surface Design of Yeast

For cell surface adsorption of rare metals by cell surface engineering, biomolecules with the ability to bind rare metals are required. Among the trace metals including rare metals, there are essential

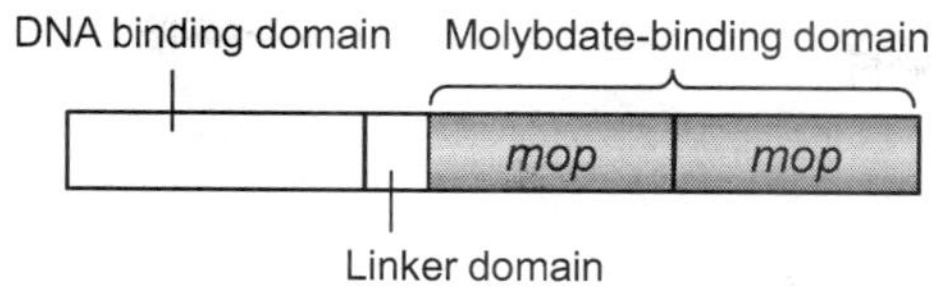

Figure 16.2. Domain structure of ModE protein.

trace metals that function as a cofactor of metalloproteins and are absolutely necessary for various enzyme reactions. Metalloproteins function normally as a result of specific recognition and binding of essential trace metals. Thus, the ability of metalloproteins to recognize and bind metals can be utilized in cell surface design for rare metal recovery.[5]

Molybdenum is one of the rare metals, and is often used in steel alloy, in electronic components, as a catalyst, and as a lubricant agent owing to its strong hardness and heat resistance. Therefore, molybdenum is an increasingly useful metal in industries, and the construction of a system for adsorbing and recovering molybdenum is required. Among the essential elements in living organisms, molybdenum plays an important role in a redox active center of the molybdopterin cofactor of many enzymes.[7] Although most molybdenum compounds are less soluble, the soluble molybdate MoO_4^{2-} is formed upon contact with oxygen and is bioavailable. In *Escherichia coli*, molybdate is transported by ABC transporters that consist of ModA (periplasmic binding protein), ModB (membrane protein), and ModC (ATPase).[8] The expression of these proteins is regulated by the transcription factor ModE. This protein regulates many aspects of molybdenum metabolism by binding to specific DNA sequences in a molybdate-binding-dependent manner.[9,10] The ModE protein has four domains, namely, N-terminal DNA-binding domain, linker region, and C-terminal molybdate-binding domain including two mop domains (Fig. 16.2).[11]

For the molecular breeding of a bioadsorbent with molybdate-binding ability, the cell surface display of ModE was performed using an α-agglutinin-based display system.[5] In this case, the C-terminal domain of ModE also was displayed on the yeast cell surface in addition to the full-length ModE (Fig. 16.3). Using the constructed cell-surface-engineered yeast displaying both types of ModE, molybdate

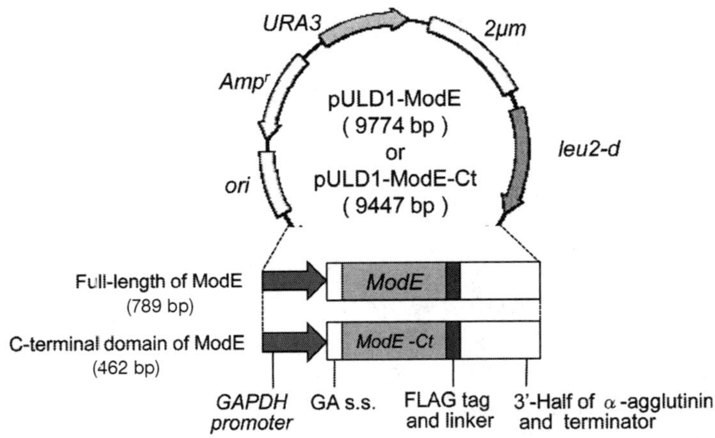

Figure 16.3. Constructed plasmids for cell surface display of full-length ModE and C-terminal domain of ModE.

adsorption in 100 μM molybdate solution was performed. After the adsorption, the residual molybdate in the solution was quantified by inductively coupled plasma mass spectrometry. As a result, the cell-surface-engineered yeast showed molybdate adsorption, indicating that the displayed full-length ModE and C-terminal domain of ModE conferred the adsorption ability on the cell surface (Fig. 16.4). The display of the C-terminal domain of ModE was more effective than that of full-length ModE in molybdate adsorption. Importantly, the cell-surface-engineered yeast showed specific binding to molybdate, except for tungstate, a molybdate analog. Furthermore, the pH of molybdate solution influenced the adsorption efficiency of the cell-surface-engineered yeast. For the next step, desorption of the adsorbed molybdate from the cell-surface-engineered yeast is an important process in terms of resource recovery. After the molybdate adsorption, yeast cells displaying the C-terminal domain of ModE were washed with buffers of various pHs, detergents, and denaturing agents, and a small amount of molybdate was recovered. However, treatment with papain allowed the recovery of molybdate, and more than 50% of the adsorbed molybdate was recovered by papain treatment in phosphate-buffered saline (pH 7.4). Papain-treated cells could be recultivated and reutilized for molybdate adsorption. Therefore, the

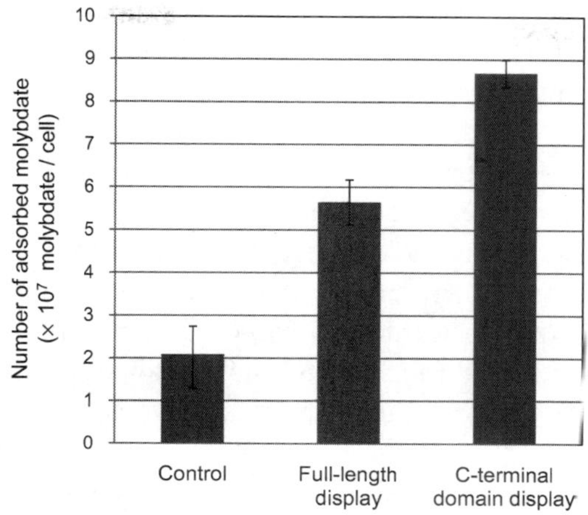

Figure 16.4. Adsorption of molybdate by cell-surface-engineered yeast in 100 μM molybdate solution (pH 3.0).

adsorption and recovery system realized by the cell surface design of yeast is useful for the recovery of valuable metal resources such as rare metals.

16.5 Further Potential of Cell Surface Design for Recovery of Rare Metal Ions

In practical adsorption and recovery of rare metal ions from the hydrosphere, there are many cases in which metal ions other than the target rare metal ions are mixed. Therefore, the construction of bioadsorbents with the ability to adsorb only the target rare metal ions selectively is the next important challenge. In the construction of bioadsorbents by cell surface design, the metal-binding function of displayed proteins is the determining factor for the metal-binding ability of bioadsorbents. Creation of proteins with selective adsorption ability may lead to production of bioadsorbents that selectively adsorb and recover rare metal ions. One of the advantages of proteins as biomolecules for specific and selective rare metal adsorption is flexibility in the alteration of proteins. Here, for the alteration of proteins by genetic engineering, intracellular and extracellular expressions (secretion) are the major procedures that convert genomic information into proteins. In an intracellular expression system, the accumulation of produced proteins inside cells could cause cellular toxicity and the formation of inactive inclusion bodies (protein aggregates). In the case of inclusion bodies, the regeneration of the protein structure (refolding) is required. On the other hand, in an extracellular expression system, proteases must be inhibited by adding protease inhibitors during the concentration procedure. These are the reasons for the decreased efficiency of protein purification. Thus, a high-throughput technology for exhaustive production and functional evaluation of proteins are desired.

Molecular display on the yeast cell surface is also an innovative molecular tool by which the function of a displayed protein can be analyzed on intact cells because various proteins such as mutated proteins can be analyzed without protein purification and concentration, which are required in the conventional methods (Fig. 16.5).[12] That is, cell-surface-engineered yeast cells can be treated as microparticles covered with proteins. This allows high-throughput screening of protein libraries for the acquisition of proteins with desirable functions. The amino acid sequence of the resultant displayed protein can be determined by DNA

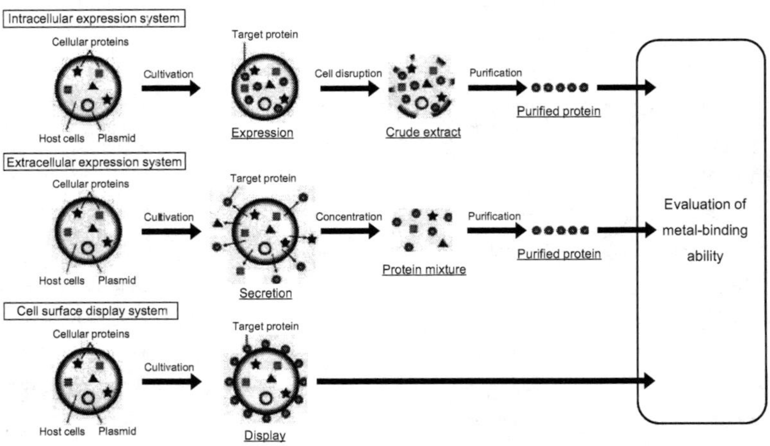

Figure 16.5. Cell surface display system as molecular tool for high-throughput screening of protein libraries. See also Color Insert.

sequencing of the introduced expression plasmids. In addition, a wide variety of useful proteins whose gene sequences are identified can be correctly folded, glycosylated, and displayed because eukaryotic yeast cells are equipped with a mechanism controlling the quality of proteins. Therefore, on the basis of cell surface engineering, cell surface design enables a universal molecular breeding of bioadsorbents. Namely, molecular breeding of bioadsorbents that specifically adsorb the target rare metal ions can be performed by altering the metal specificity of metal-binding proteins, creation of novel peptides and proteins with specific adsorption ability, and their display on the yeast cell surface. For this purpose, the libraries consisting of metal-binding proteins that have combinatorial mutations in the region for metal ion recognition and the random peptide libraries consisting of peptides with random sequences of amino acids are constructed as a first step. Then, yeast libraries are constructed by the cell surface display of the protein and peptide libraries. By the screening of proteins with specific adsorption ability from the yeast libraries, the construction of novel metal-binding biomolecules and bioadsorbents is promising in further studies (Fig. 16.6).

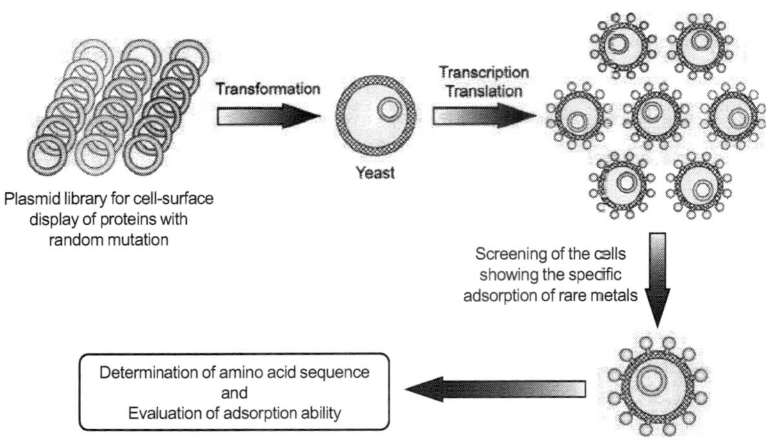

Figure 16.6. Creation of bioadsorbents that selectively adsorb rare metal ions by screening yeast libraries displaying peptides or proteins with combinatorial mutations. See also Color Insert.

16.6 Conclusion

The molecular breeding of a resource-recycling bioadsorbent was achieved by the cell surface adsorption approach and the establishment of cell surface engineering that enables cell surface design. As described in this chapter, the cell-surface-engineered yeast displaying ModE was effective for the adsorption and recovery of molybdate, as a resource-recycling bioadsorbent. A cell surface adsorption system is important in modern industrial societies because an efficient reutilization of rare metals is required in various fields such as high-tech industries. By replacing proteins displayed on the yeast cell surface according to the target metal ions, a bioadsorbent can be custom-made. Molecular display on the yeast cell surface is also useful in high-throughput evaluation of protein functions. Using this technology, the creation of novel biomolecules accommodating a wide range of rare metal and rare earth ions is possible, even if such biomolecules are nonexistent in the genome. The flexibility of cell surface design against diverse metal species is also an important advantage. On the basis of cell surface engineering, the creation of novel metal-binding proteins and basic analysis of metal recognition of proteins would contribute to the further development of a future-oriented resource-recycling biosystem.

References

1. Kuroda, K., Shibasaki, S., Ueda, M., and Tanaka, A., *Appl. Microbiol. Biotechnol.* **57**, 697–701 (2001).
2. Kuroda, K., Ueda, M., Shibasaki, S., and Tanaka, A., *Appl. Microbiol. Biotechnol.* **59**, 259–264 (2002).
3. Kuroda, K. and Ueda, M., *Appl. Microbiol. Biotechnol.* **63**, 182–186 (2003).
4. Kuroda, K. and Ueda, M., *Appl. Microbiol. Biotechnol.* **70**, 458–463 (2006).
5. Nishitani, T., Shimada, M., Kuroda, K., and Ueda, M., *Appl. Microbiol. Biotechnol.* in press.
6. Walker, S. G., Flemming, C. A., Ferris, F. G., Beveridge, T. J., and Bailey, G. W., *Appl. Environ. Microbiol.* **55**, 2976–2984 (1989).

7. Kisker, C., Schindelin, H., and Rees, D. C., *Annu. Rev. Biochem.* **66**, 233–267 (1997).
8. Self, W. T., Grunden, A. M., Hasona, A., and Shanmugam, K. T., *Res. Microbiol.* **152**, 311–321 (2001).
9. McNicholas, P. M., Mazzotta, M. M., Rech, S. A., and Gunsalus, R. P., *J. Bacteriol.* **180**, 4638–4643 (1998).
10. Hollenstein, K., Frei, D. C., and Locher, K. P., *Nature* **446**, 213–216 (2007).
11. Wagner, U. G., Stupperich, E., and Kratky, C., *Structure* **8**, 1127–1136 (2000).
12. Ueda, M., *J. Mol. Catal. B* **28**, 139–143 (2004).

Chapter 17

BIOINFORMATICS TOOLS FOR THE NEXT GENERATION OF METAL BIOTECHNOLOGY

Hideki Nakayama
Faculty of Environmental Studies, Nagasaki University, 1-14 Bunkyo-machi, Nagasaki 852-8521, Japan
nakayamah@nagasaki-u.ac.jp

17.1 Introduction

At present, above 5900 genomes of living organisms have been completely sequenced and stored as text-based data files through the Internet-based database systems. Most database systems were made freely available to the public through the World Wide Web-based systems, such as the National Center for Biotechnology Information (NCBI) Entrez,[1] the Kyoto Encyclopedia of Genes and Genomes (KEGG),[2] and the Genomes OnLine Database (GOLD).[3] It is estimated that more than one third of all known proteins are metalloproteins, which require metals to maintain their structures and functions in living organisms.[4] Obviously, there is not enough experimental evidence to elucidate metal-selection mechanisms of the complete set of metalloproteins encoded by genomes of living

Handbook of Metal Biotechnology: Applications for Environmental Conservation and Sustainability
Edited by Michihiko Ike, Mitsuo Yamashita, and Satoshi Soda

www.panstanford.com

organisms. Based on Irving–Williams series of relative stabilities of complexes formed by metal ions ($Ca^{2+} < Mg^{2+} < Mn^{2+} < Fe^{2+} < Co^{2+} < Ni^{2+} < Cu^{2+} > Zn^{2+}$),[5] metalloproteins would bind most strongly to divalent (cupric) copper, and to a lower strength to other metal ions. To maintain life systems, however, metalloproteins must selectively bind to specific metal ions required for their function. Recently, it is proposed that metal-binding selectivity of metalloproteins is determined by spatiotemporal folding of the proteins in the cells, and not only by the basis of the nature, number, and geometric arrangement of the binding residues, or the size and charge of the metal-binding pocket.[6]

In the field of metal biotechnology research, it is important to discover and engineer metalloproteins with improved functions involving recognition or sensing of metals, chelating or binding of metals, and reduction or oxidation of metals, which are useful for various types of metal biotechnologies, such as metal-biosensors for monitoring of metal-pollution, bioleaching of metals from their ores, and either bioadsorption or biomineralization for recovering ionic metals from environmental water systems, as shown in other chapters of this book. However, discovery of novel and superior metalloproteins is still difficult due to a lack of reliable high-throughput experimental procedures. Therefore, bioinformatics tools are required as a primary screening tool to predict and identify a group of candidate metalloproteins for further identification and confirmation. For example, based on nucleotide and amino acid sequences, putative metalloproteins, such as zinc-finger proteins or metalloenzymes, can be predicted by the structural similarity with known metalloproteins, the presence of specific metal-binding sites, or metal-binding domains. Subsequently, the predicted metalloproteins can be analyzed for their functional relationship with specific metals in the cells, and the superior metalloproteins identified by this process can be used for development of metal biotechnology.

17.2 Public Protein Database as a Gold Mine

As the number of completely sequenced genomes increases in public databases, international research communities are now refocusing on collecting information about all the proteins encoded in these genomes. Emerging public protein databases allow us to access and rationalize large amount of protein data.

17.2.1 *Universal Protein Resource*

The Universal Protein Resource (UniProt; http://www.uniprot.org/) is a comprehensive resource for protein sequence and annotation data.[7] UniProt is produced by the UniProt Consortium which consists of groups from the European Bioinformatics Institute (EBI), the Swiss Institute of Bioinformatics (SIB), and the Protein Information Resource (PIR). UniProt is composed of four major components, each optimized for different uses: the UniProt Archive (UniParc), the UniProt Knowledgebase (UniProtKB), the UniProt Reference Clusters (UniRef), and the UniProt Metagenomic and Environmental Sequence Database (UniMES), as follows.

(a) *UniParc* is the most comprehensive publicly accessible non-redundant protein sequence database available, providing links to all underlying sources and versions of these sequences. Researchers can instantly find out whether a sequence of interest is already in the public domain and, if not, identify its closest relatives.
(b) *UniProtKB* is used to access functional information on proteins. The UniProtKB consists of two sections: Swiss-Prot, which is manually annotated and reviewed, and TrEMBL, which is automatically annotated and is not reviewed. Every UniProtKB entry contains the amino acid sequence, protein name or description, taxonomic data, and citation information. In addition, UniProtKB contains further annotation that includes widely accepted biological ontologies, classifications and cross-references, as well as clear indications on the quality of annotation in the form of evidence attribution to experimental and computational data.
(c) *UniRef* provides clustered sets of sequences from UniProtKB and selected UniParc records. UniRef90 and UniRef50 yield a database size reduction of approximately 40% and

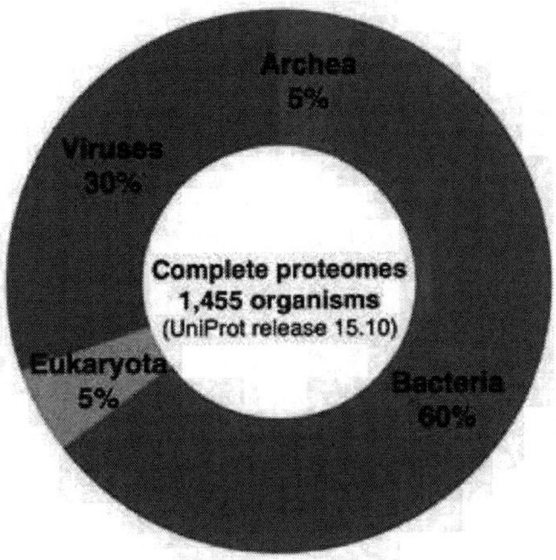

Figure 17.1. Distributions of organisms in the complete proteomes of UniProt. To be included in the complete proteomes, an organism must have a completely sequenced genome, i.e., fully closed and exhibiting either good gene prediction models or good quality transcriptome/proteome data. Therefore, for bacterial and archaeal genomes, whole-genome shotguns (WGS) and draft sequences are not included. See also Color Insert.

65%, respectively, providing significantly faster sequence searches.

(d) *UniMES* is a repository specifically for metagenomic and environmental data.

At present, UniProt (release 15.10) contains complete proteomes of total 1455 organisms including 68 archea, 872 bacteria, 76 eukariota, and 439 viruses (Fig. 17.1).

Searching UniProtKB under the field of Gene Ontology (GO) for entry whose molecular function is metal ion binding (GO:0046872), a total of 772,892 entries are retrieved. These entries are considered to be putative metalloproteins. The detail results for each metal ion are shown in Table 17.1. Currently, 16 types of metalloproteins are found based on their possible binding to 16 different types of metals including 3 alkali metals (Li, Na, and K), 2 alkaline earth metals (Mg and Ca), 10 transition metals (V, Mn, Fe, Co, Ni, Cu, Zn, Mo, Cd, Hg), and 1 other metal (Pb).

Table 17.1. Sixteen types of putative metalloproteins listed in UniProtKB.

Metal ions	Number of metalloproteins	GO IDs
Iron (Fe) ion	324,953	GO:0005506
Zinc (Zn) ion	229,453	GO:0008270
Copper (Cu) ion	117,795	GO:0005507
Magnesium (Mg) ion	79,617	GO:0000287
Calcium (Ca) ion	37,669	GO:0005509
Manganese (Mn) ion	23,548	GO:0030145
Potassium (K) ion	9,597	GO:0030955
Molybdenum (Mo) ion	9,444	GO:0030151
Nickel (Ni) ion	8,038	GO:0016151
Cobalt (Co) ion	6,469	GO:0050897
Sodium (Na) ion	4,028	GO:0031402
Mercury (Hg) ion	660	GO:0045340
Cadmium (Cd) ion	132	GO:0046870
Lithium (Li) ion	36	GO:0031403
Vanadium (V) ion	18	GO:0051212
Lead (Pb) ion	2	GO:0032791

17.2.2 *Worldwide Protein Data Bank*

The Worldwide Protein Data Bank (wwPDB; http://www.wwpdb.org/) consists of organizations that act as deposition, data processing, and distribution centers for PDB data.[8] The members are RCSB PDB (USA), PDBe (Europe), and PDBj (Japan). The mission of the wwPDB is to maintain a single PDB Archive of macromolecular structural data that is freely and publicly available to the global community. Currently, total number of wwPDB depositions is more than 56,000 structures.

In addition, the chemical component dictionary is found in wwPDB as an external reference file describing all residue and small molecule components. This dictionary contains detailed chemical descriptions for standard and modified amino acids/nucleotides, small molecule ligands, and solvent molecules, which include metal ions and metal-containing ligand molecules. The PDBeChem (http://www.ebi.ac.uk/msd-srv/chempdb/), one of the PDBe services, offers possibilities for searching and exploring the dictionary. The chemical component dictionary currently contains 224 different

Table 17.2. Metal-containing ligand molecules in PDB entries.

Core metals	Number of metal ions and metal-containing ligand molecules	Number of linked PDB entries
Fe	54	1,759
Cu	22	717
Mo	19	111
Ni	16	426
Hg	16	395
Zn	13	5,320
Mg	13	4,901
Co	13	459
Mn	12	1,327
Ca	11	4,671
Na	11	2,457
V	10	64
K	5	867
Cd	4	484
Pb	3	40
Li	2	34
Total	224	24,032

types of metal ions and metal-containing ligand molecules, which are linked to 24,032 PDB entries (Table 17.2).

17.3 Mining of Protein Domains

Proteins are generally composed of one or more core functional elements, commonly termed domains. Different combinations of domains allow evolving the diverse range of proteins with various functions in nature. The identification of domains that occur within proteins can provide valuable insights into their functions. As metal-binding domains are core functional elements found in metalloproteins, the bioinformatics tools for mining of metal-binding domains would be advantageous to all researchers in the field of metal biotechnology.

17.3.1 *InterPro: The Integrative Protein Signature Database*

InterPro (http://www.ebi.ac.uk/interpro/) is a database of protein families, domains, regions, repeats, sites, and posttranslational modifications (PTMs) of known proteins.[9] InterPro (release 23.1) contains 19,269 entries, representing 82 active sites, 58 binding sites, 546 conserved sites, 5466 domains, 11,537 families, 1302 regions, 23 PTMs, and 255 repeats.

17.3.1.1 Protein family and domain databases integrated with InterPro

Now InterPro is integrated with member databases including UniProtKB and other 11 protein family and domain databases shown in Table 17.3.

(a) *Pfam* is a comprehensive collection of protein domains and families, represented as multiple sequence alignments and as profile Hidden Markov Models (HMMs).[10] The Pfam (release 24.0) contains 11,912 protein families. The Pfam is

Table 17.3. Members of protein-family and domain database integrated in InterPro (release 23.1) consortium.

Member database	Integrated version	Signatures*	Integrated signatures**
Pfam	23.0	10,340	10,329
TIGRFAMs	8.0	3,603	3,581
PIRSF	2.70	2,742	2,691
PANTHER	6.1	30,128	2,234
PROSITE	20.52	2,168	2,141
PRINTS	39.0	1,950	1,927
SUPERFAMILY	1.69	1,538	1,090
Gene3D	3.0.0	2,147	1,026
ProDom	2006.1	1,894	992
SMART	6.0	809	804
HAMAP	280509	1,633	502
UniProtKB	15.10	8,123,918	7,813,392

*Some signatures may not have matches to UniProtKB proteins.
**Not all signatures of a member database may be integrated at the time of InterPro release.

available on the web from the consortium members using the web sites in the UK (http://pfam.sanger.ac.uk/), the USA (http://pfam.janelia.org/), and Sweden (http://pfam.sbc.su.se/), as well as from mirror sites in France (http://pfam.jouy.inra.fr/) and South Korea (http://pfam.ccbb.re.kr/).

(b) *TIGRFAMs* (http://www.jcvi.org/cms/research/projects/tigrfams/) are a collection of protein families featuring curated multiple sequence alignments, HMMs, and associated information designed to support the automated functional identification of proteins by sequence homology.[11]

(c) *PIRSF* (Protein Information Resource SuperFamily) classification system (http://pir.georgetown.edu/pirsf/) reflects evolutionary relationships of full-length proteins and domains.[12] The primary PIRSF classification unit is the homeomorphic family, whose members are both homologous (evolved from a common ancestor) and homeomorphic (sharing full-length sequence similarity and a common domain architecture). PIRSF families are curated systematically based on literature review and integrative sequence and functional analysis, including sequence and structure similarity, domain architecture, functional association, genome context, and phyletic pattern.

(d) *PANTHER* (Protein ANalysis THrough Evolutionary Relationships) classification system (http://www.pantherdb.org/) classifies genes by their functions, using published scientific experimental evidence and evolutionary relationships to predict function even in the absence of direct experimental evidence.[13] Proteins are classified by expert biologists into families and subfamilies of shared function, which are then categorized by molecular function and biological process ontology terms. For an increasing number of proteins, detailed biochemical interactions in canonical pathways are captured and can be viewed interactively.

(e) *PROSITE* (http://www.expasy.org/prosite/) is a protein domain database for functional characterization and annotation.[14] The PROSITE (release 20.54) contains 1308 patterns, 863 profiles, and 869 ProRules.

(f) *PRINTS* (http://www.bioinf.manchester.ac.uk/dbbrowser/PRINTS/) is a compendium of protein fingerprints.[15] A fingerprint is a group of conserved motifs used to characterize a protein family; its diagnostic power is refined by iterative scanning of a SWISS-PROT/TrEMBL composite. Usually the motifs do not overlap, but are separated along a sequence, though they may be contiguous in 3D-space. Fingerprints can encode protein folds and functionalities more flexibly and powerfully than can single motifs, full diagnostic potency deriving from the mutual context provided by motif neighbors.

(g) *SUPERFAMILY* (http://supfam.cs.bris.ac.uk/SUPERFAMILY/) is a database of structural and functional annotation for all proteins and genomes.[16] The SUPERFAMILY annotation is based on a collection of HMMs, which represent structural protein domains at the Structural Classification of Proteins (SCOP)[17] superfamily level. The annotation is produced by scanning protein sequences from over 1200 completely sequenced genomes against the HMMs.

(h) *Gene3D* (http://gene3d.biochem.ucl.ac.uk/) provides accurate structural domain family assignments for over 1100 genomes and nearly 10 million proteins.[18] A HMM library, constructed from the manually curated CATH (Class, Architecture, Topology, Homology) structural domain hierarchy,[19] is used to search UniProt, RefSeq, and Ensembl[20] protein sequences. The resulting matches are refined into simple multi-domain architectures using a recently developed algorithm, DomainFinder 3 (ftp://ftp.biochem.ucl.ac.uk/pub/gene3d_data/DomainFinder3/). The domain assignments are integrated with multiple external protein function descriptions (e.g. GO and KEGG), structural annotations (e.g. coiled coils, disordered regions, and sequence polymorphisms) and family resources (e.g. Pfam and eggNog[21]) and displayed on the Gene3D website.

(i) *ProDom* (http://prodom.prabi.fr/) is a comprehensive set of protein domain families automatically generated from the UniProtKB.[22] The ProDom (release 2006.1) contains 1,716,114 domain families.

(j) *SMART* (Simple Modular Architecture Research Tool; http://smart.embl.de/) is an online tool for the identification and annotation of protein domains.[23] It provides a user-friendly platform for the exploration and comparative study of domain architectures in both proteins and genes. The SMART (release 6.0) contains manually curated models for 784 protein domains. The underlying protein database is based on completely sequenced genomes of 630 species. The interaction network view is available for more than 2 million proteins.

(k) *HAMAP* (High-quality Automated and Manual Annotation of microbial Proteomes) system (http://www.expasy.org/sprot/hamap) is composed of two databases, the proteome database and the family database.[24] The proteome database comprises biological and sequence information for each completely sequenced microbial proteome. The family database currently comprises more than 1600 manually curated orthologous protein families that belong to one of the HAMAP families.

17.3.1.2 Functional molecules for metal ion binding found in InterPro

Searching InterPro under the field of GO for entry whose molecular function is metal ion binding (GO:0046872), a total of 789 entries are found as functional molecules for metal-ion binding among known proteins. The results of each GO ID are extracted in Table 17.4. Among of the known metalloproteins, protein families or domains binding to Zn or Fe ions are well-characterized at present, while no entries are found for protein families or domains binding to Li or Pb ions in InterPro.

17.3.2 *Bioinformatics Tools for Prediction of Metal-Binding Domains*

Since number of sequences and structures of proteins with unknown biological function are continually accumulating in public databases, sophisticated and efficient tools for metal-binding domain

Table 17.4. Numbers of functional molecules as metal ion binding in InterPro.

Metals	Number of InterPro entry	GO IDs
Zn	228	GO:0008270
Fe	129	GO:0005506
Ca	88	GO:0005509
Mg	83	GO:0000287
Cu	47	GO:0005507
Ni	28	GO:0016151
Mn	18	GO:0030145
Mo	14	GO:0030151
Co	6	GO:0050897
K	4	GO:0030955
Hg	3	GO:0045340
Na	2	GO:0031402
Cd	1	GO:0046870
V	1	GO:0051212
Li	0	GO:0031403
Pb	0	GO:0032791

prediction are needed for further progress of metal biotechnology. At present, several publicly available bioinformatics tools have been developed to predict metal binding residues from sequence data. These tools, listed below, can be used for finding and design novel metal-binding domains.

(a) *MDB* (Metalloprotein Database and Browser; http:// metallo. scripps.edu/) is a web-accessible resource for metalloprotein research.[25] It includes quantitative information on geometrical parameters of metal-binding sites in protein structures available from the wwPDB.
(b) *MetSite* (http://bioinf.cs.ucl.ac.uk/MetSite/) represents a fully automatic approach for the detection of metal-binding residue clusters applicable to protein models of moderate quality.[26] The method involves using sequence profile information in combination with approximate structural data. MetSite allows users to scan query structures using one of the six metal type (Ca, Zn, Mg, Fe, Cu, and Mn) classifiers.
(c) *FEATURE metal scanning* (http://feature.stanford.edu/ metals/) is a currently developing tool for identification of

metal binding sites in proteins with no exist sequence similarity to known structures.[27] At present, only zinc binding sites could be identified by this tool.

(d) *MetalDetector* (http://metaldetector.dsi.unifi.it/) is a classifier that predicts transition-metal binding for Cys and His residues in protein; for Cys it also predicts disulfide bonding bridges.[28]

(e) *CHED server* (http://ligin.weizmann.ac.il/ched/) uses the "CHED" algorithm to predict 3D intra-chain protein binding sites for transition metals (Zn, Fe, Mn, Cu, Ni, Co), and for Ca and Mg sites that can be replaced by a transition metal.[29] The algorithm searches for a triad of amino acids composed of four residue types (Cys, His, Glu, Asp; CHED) having ligand atoms within specific distances.

(f) *SeqCHED server* (http://ligin.weizmann.ac.il/seqched/) is a sequence-based prediction server that enables the user to analyze a translated gene sequence for transition metals (Zn, Fe, Ni, Cu, Co, Mn), and for Ca and Mg binding sites.[30] The application checks for homology of your target sequence to PDB template sequences and then models the target side chains in 3D (using SCCOMP[31]) on the backbone of the selected template. A metal binding prediction algorithm (based on the CHED procedure) is then applied to the 3D model to identify any putative binding sites and their ligating CHED residues.

17.4 The Next Generation of Metal Biotechnology

Using the current bioinformatics tools, researchers could obtain lists of putative metalloproteins and metal-binding domains. However, for further innovation for next generation of metal biotechnology, we need to develop and establish high-throughput experimental methods for analyzing the function of metalloproteins and metal-binding domains on those lists.

Recent progress on ionome,[32,33] metalloproteomes,[34] and metallomics[35] are brought on by the use of efficient high-throughput analytical machineries and sophisticated bioinformatics tools.

Researchers are focusing now on the continually accumulating data obtained from these studies to elucidate interactions between metal ions and biomolecules including genome, proteome, and metabolome. These omics studies would clarify the role of metal ions, metalloproteins, and metal-binding domains in living organisms and provide crucial ideas for innovation of the next generation of metal biotechnology. Alternatively, combinatorial bioengineering integrated with cell-surface display system[36] would be a promising approach to create and screen novel function of metalloproteins and metal-binding domains. Although it is still challenging, functional design of metalloproteins has been initiated.[37] Further progression of both metalloproteins and bioinformatics studies would open gateway to the future of metal biotechnology.

References

1. Baxevanis, A. D., Searching NCBI databases using Entrez. *Curr. Protoc. Bioinformatics*, Chapter 1, pp. Unit 1 3 (2008).
2. Aoki-Kinoshita, K. F. and Kanehisa, M., *Methods Mol. Biol.* **396**, 71–91 (2007).
3. Bernal, A., Ear, U., and Kyrpides, N., *Nucleic Acids Res.* **29**, 126–127 (2001).
4. Tainer, J. A., Roberts, V. A., and Getzoff, E. D., *Curr. Opin. Biotechnol.* **3**, 378–387 (1992).
5. Irving, H. and Williams, R. J. P., *Nature* **162**, 746–747 (1948).
6. Tottey, S., Waldron, K. J., Firbank, S. J., Reale, B., Bessant, C., Sato, K., Cheek, T. R., Gray, J., Banfield, M. J., Dennison, C., and Robinson, N. J., *Nature* **455**, 1138–1142 (2008).
7. The UniProt Consortium, *Nucleic Acids Res.* (2009).
8. Berman, H., Henrick, K., Nakamura, H., and Markley, J. L., *Nucleic Acids Res.* **35**, D301–D303 (2007).
9. Hunter, S., Apweiler, R., Attwood, T. K., Bairoch, A., Bateman, A., Binns, D., Bork, P., Das, U., Daugherty, L., Duquenne, L., Finn, R. D., Gough, J., Haft, D., Hulo, N., Kahn, D., Kelly, E., Laugraud, A., Letunic, I., Lonsdale, D., Lopez, R., Madera, M., Maslen, J., McAnulla, C., McDowall, J., Mistry, J., Mitchell, A., Mulder, N., Natale, D., Orengo, C., Quinn, A. F., Selengut, J. D., Sigrist,

C. J., Thimma, M., Thomas, P. D., Valentin, F., Wilson, D., Wu, C. H., and Yeats, C., *Nucleic Acids Res.* **37**, D211–D215 (2009).

10. Finn, R. D., Tate, J., Mistry, J., Coggill, P. C., Sammut, S. J., Hotz, H. R., Ceric, G., Forslund, K., Eddy, S. R., Sonnhammer, E. L., and Bateman, A., *Nucleic Acids Res.* **36**, D281–D288 (2008).
11. Haft, D. H., Selengut, J. D., and White, O., *Nucleic Acids Res.* **31**, 371–373 (2003).
12. Nikolskaya, A. N., Arighi, C. N., Huang, H., Barker, W. C., and Wu, C. H., *Evol. Bioinform. Online* **2**, 197–209 (2006).
13. Mi, H., Guo, N., Kejariwal, A., and Thomas, P. D., *Nucleic Acids Res.* **35**, D247–D252 (2007).
14. Sigrist, C. J., Cerutti, L., de Castro, E., Langendijk-Genevaux, P. S., Bulliard, V., Bairoch, A., and Hulo, N., *Nucleic Acids Res.* (2009).
15. Attwood, T. K., Beck, M. E., Flower, D. R., Scordis, P., and Selley, J. N., *Nucleic Acids Res.* **26**, 304–308 (1998).
16. Wilson, D., Madera, M., Vogel, C., Chothia, C., and Gough, J., *Nucleic Acids Res.* **35**, D308–D313 (2007).
17. Andreeva, A., Howorth, D., Chandonia, J. M., Brenner, S. E., Hubbard, T. J., Chothia, C., and Murzin, A. G., *Nucleic Acids Res.* **36**, D419–D425 (2008).
18. Lees, J., Yeats, C., Redfern, O., Clegg, A., and Orengo, C., *Nucleic Acids Res.* (2009).
19. Cuff, A. L., Sillitoe, I., Lewis, T., Redfern, O. C., Garratt, R., Thornton, J., and Orengo, C. A., *Nucleic Acids Res.* **37**, D310–D314 (2009).
20. Hubbard, T. J., Aken, B. L., Ayling, S., Ballester, B., Beal, K., Bragin, E., Brent, S., Chen, Y., Clapham, P., Clarke, L., Coates, G., Fairley, S., Fitzgerald, S., Fernandez-Banet, J., Gordon, L., Graf, S., Haider, S., Hammond, M., Holland, R., Howe, K., Jenkinson, A., Johnson, N., Kahari, A., Keefe, D., Keenan, S., Kinsella, R., Kokocinski, F., Kulesha, E., Lawson, D., Longden, I., Megy, K., Meidl, P., Overduin, B., Parker, A., Pritchard, B., Rios, D., Schuster, M., Slater, G., Smedley, D., Spooner, W., Spudich, G., Trevanion, S., Vilella, A., Vogel, J., White, S., Wilder, S., Zadissa, A., Birney, E., Cunningham, F., Curwen, V., Durbin, R., Fernandez-Suarez, X. M., Herrero, J., Kasprzyk, A., Proctor, G., Smith, J., Searle, S., and Flicek, P., *Nucleic Acids Res.* **37**, D690–D697 (2009).
21. Muller, J., Szklarczyk, D., Julien, P., Letunic, I., Roth, A., Kuhn, M., Powell, S., von Mering, C., Doerks, T., Jensen, L. J., and Bork, P., *Nucleic Acids Res.* (2009).

22. Bru, C., Courcelle, E., Carrere, S., Beausse, Y., Dalmar, S., and Kahn, D., *Nucleic Acids Res.* **33**, D212–D215 (2005).
23. Letunic, I., Doerks, T., and Bork, P., *Nucleic Acids Res.* **37**, D229–D232 (2009).
24. Lima, T., Auchincloss, A. H., Coudert, E., Keller, G., Michoud, K., Rivoire, C., Bulliard, V., de Castro, E., Lachaize, C., Baratin, D., Phan, I., Bougueleret, L., and Bairoch, A., *Nucleic Acids Res.* **37**, D471–D478 (2009).
25. Castagnetto, J. M., Hennessy, S. W., Roberts, V. A., Getzoff, E. D., Tainer, J. A., and Pique, M. E., *Nucleic Acids Res.* **30**, 379–382 (2002)
26. Sodhi, J. S., Bryson, K., McGuffin, L. J., Ward, J. J., Wernisch, L., and Jones, D. T., *J. Mol. Biol.* **342**, 307–320 (2004).
27. Ebert, J. C. and Altman, R. B., *Protein Sci.* **17**, 54–65 (2008).
28. Lippi, M., Passerini, A., Punta, M., Rost, B., and Frasconi, P., *Bioinformatics* **24**, 2094–2095 (2008).
29. Babor, M., Gerzon, S., Raveh, B., Sobolev, V., and Edelman, M., *Proteins* **70**, 208–217 (2008).
30. Levy, R., Edelman, M., and Sobolev, V., *Proteins* **76**, 365–374 (2009).
31. Eyal, E., Najmanovich, R., McConkey, B. J., Edelman, M., and Sobolev, V., *J. Comput. Chem.* **25**, 712–724 (2004).
32. Williams, L. and Salt, D. E., *Curr. Opin. Plant Biol.* **12**, 247–249 (2009).
33. Eide, D. J., Clark, S., Nair, T. M., Gehl, M., Gribskov, M., Guerinot, M. L., and Harper, J. F., *Genome Biol.* **6**, R77 (2005).
34. Andreini, C., Bertini, I., and Rosato, A., *Acc. Chem. Res.* (2009).
35. Mounicou, S., Szpunar, J., and Lobinski, R., *Chem. Soc. Rev.* **38**, 1119–1138 (2009).
36. Kondo, A. and Ueda, M., *Appl. Microbiol. Biotechnol.* **64**, 28–40 (2004).
37. Lu, Y., Yeung, N., Sieracki, N., and Marshall, N. M., *Nature* **460**, 855–862 (2009).

Index

Color Insert

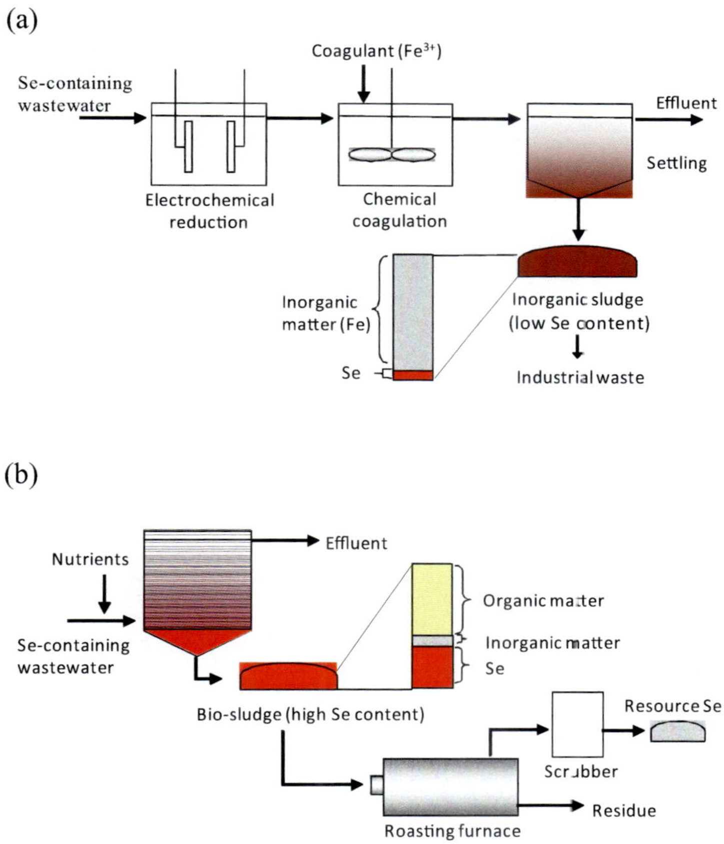

Figure 3.4. Schematic treatment systems for selenium-containing wastewater. (a) A typical physico-chemical system using an electrochemical reduction process and a coagulation process, (b) a new system using a biological reduction process and a roasting furnace for selenium recovery.

Figure 6.3. Effect of chelating agents on the growth of *Brassica juncea*. Photographs of *B. juncea* at the beginning of the experiment (a), after seven days without chelating agents (b), after seven days with EDTA (5 mM) (c), after seven days with EDDS (5 mM) (d), and after seven days with citrate (5 mM) (e) are shown.

Figure 7.1. *Astragulus sinicus* (Chinese milk vetch, or renge-soh in Japanese).

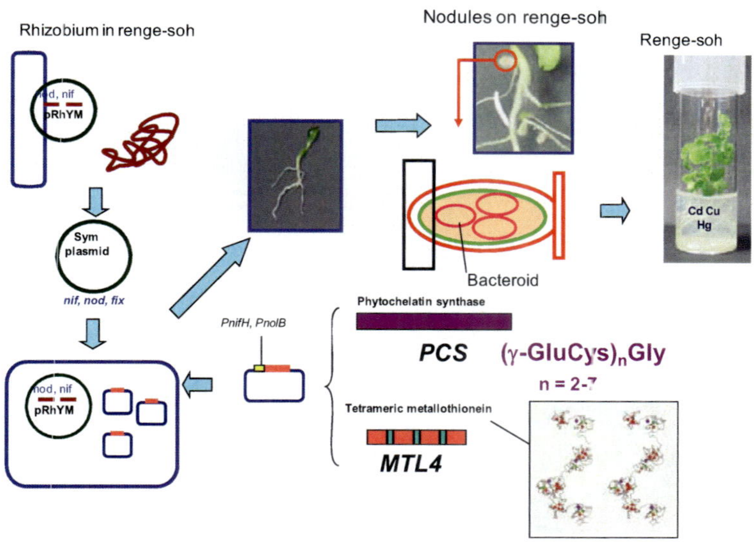

Figure 7.2. Creation of symbiosis biomaterial for metal purification.

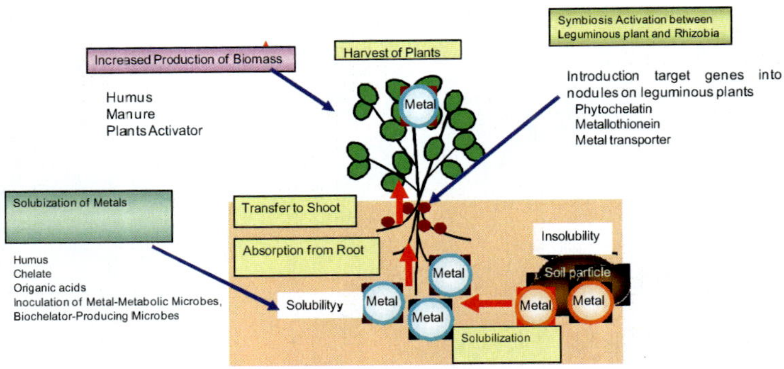

Figure 7.3. Strategy of phytoremediation technology for metal contaminants.

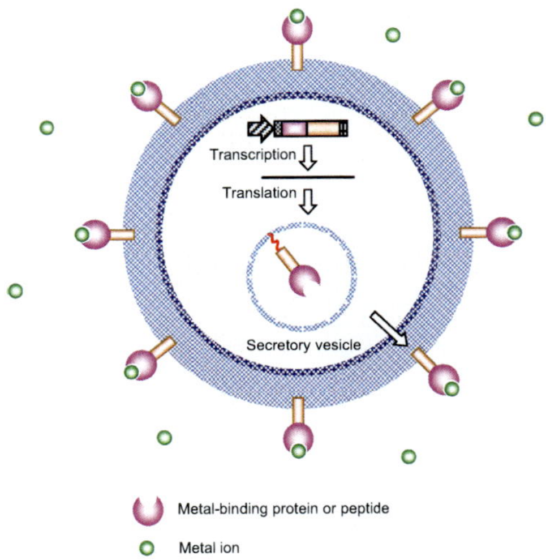

Figure 8.2. Model of yeast bioadsorbent constructed by cell surface engineering.

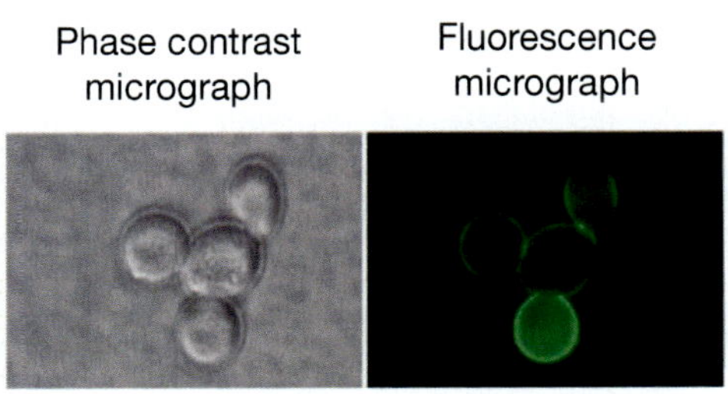

Figure 8.4. Confirmation of cell surface display of hexa-His by immunofluorescence labeling.

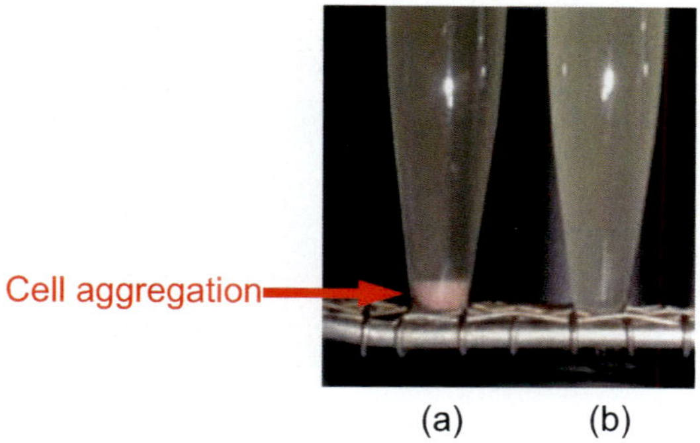

Figure 8.6. Self-aggregation of yeast cells harboring *CUP1* promoter-*GTS1* fusion gene in response to environmental copper ions. (a) Presence of 100 μM copper ions, (b) without copper ions.

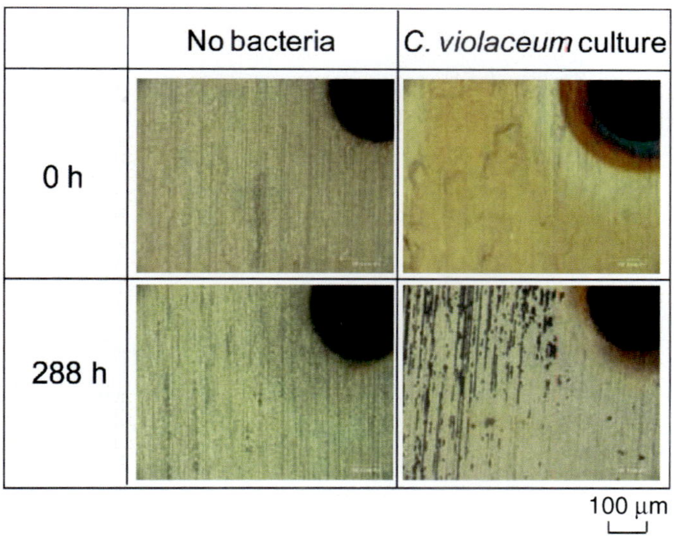

Figure 10.3. Change of surface morphology of Au plated Cu conductor on PWB immersed in culture medium with and without *C. violaceum*.

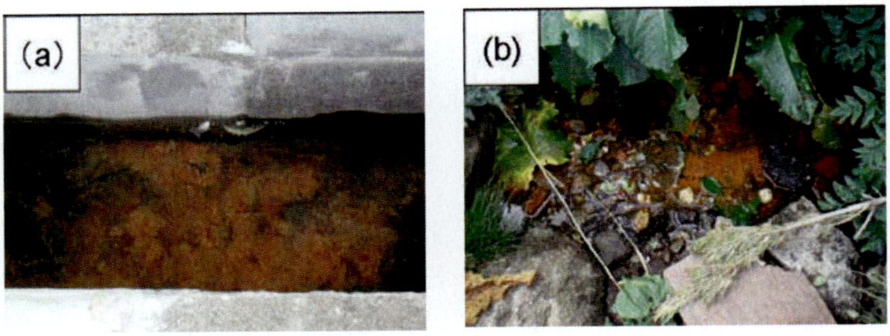

Figure 12.3. Photographs of ocherous precipitates in side ditch. (a) The side ditch in Okayama University in Japan. (b) Beside of small steam in Vitosha mountain in Bulgaria. These ocherous precipitates are colonies of iron-oxidizing bacteria and iron oxides produced by them.

Figure 12.5. Photograph of water purification plant of Joyo city in Kyoto prefecture in Japan. All ocherous precipices are iron oxides produced by iron-oxidizing bacteria.

Figure 13.1. Natural oil-producing wells and ponds in Japan, where samples were collected for the isolation and enrichment of microorganisms. a: Oil-producing wells in Niitsu, Niigata; b: a natural oil-producing pond in Singleton memorial park (Abura-tsubo: Oil vessel); c: a natural oil-producing pond in Showa, Akita; d, e, f: oil-producing wells in Showa and Akita city, Akita.

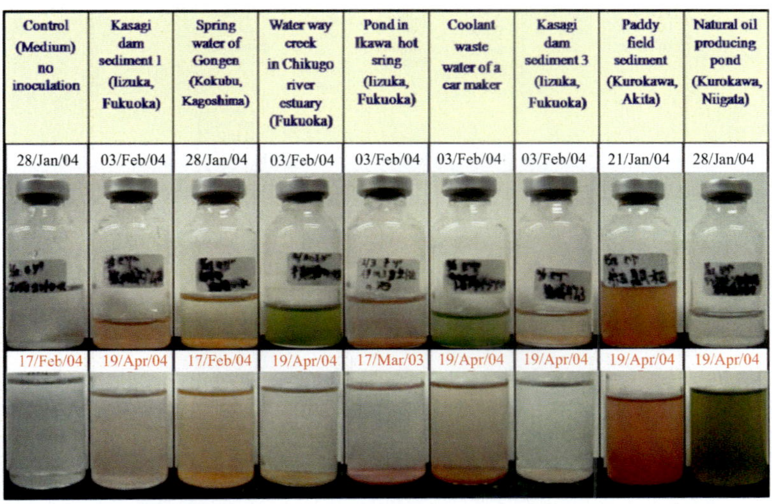

Figure 13.2. Appearance of precipitation in enrichment contained zinc and selenate (1 mM respectively) for ZnSe formation (upper line: sampling place, middle line: 1st enrichment, lower line: transferred enrichment (>2nd)).

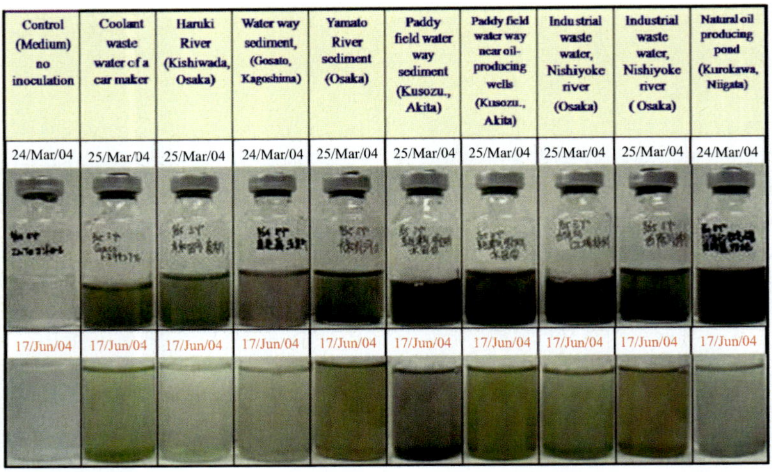

Figure 13.3. Appearance of precipitation in enrichment contained zinc and tellurite (1 mM respectively) for ZnTe formation (upper line: sampling place, middle line: 1st enrichment, lower line: transferred enrichment (>2nd)).

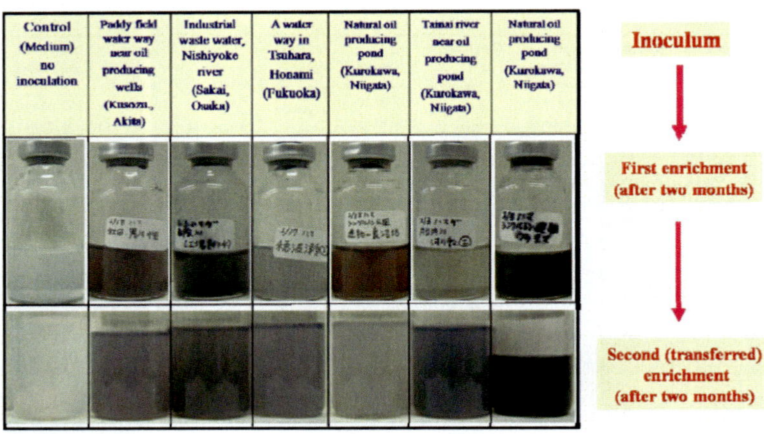

Figure 13.4. Appearance of precipitation in enrichments contained cadmium and tellurite (1 mM respectively) for CdTe formation.

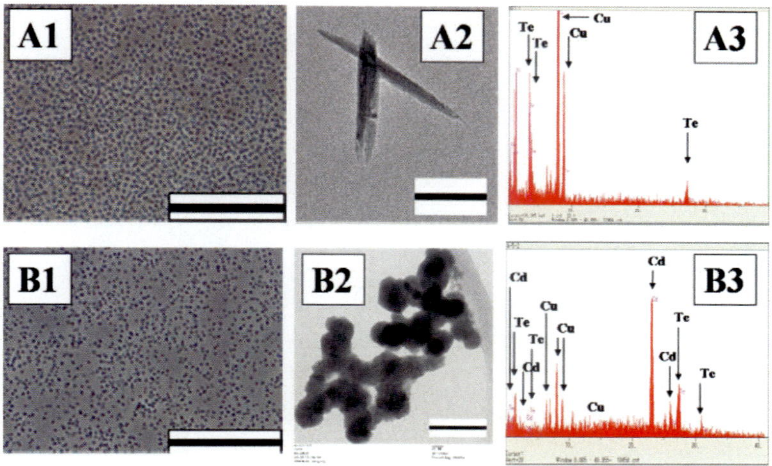

Figure 13.9. Phase-contrast photomicrographs of microbial isolate from enrichments (NT-ER and SM-ER consortia), and TEM image and EDX elemental analysis of the nano-particle produced by the isolate. a1 and b1 show phase-contrast photomicrographs of strains NTT-1 (from consortium NT-ER) and SM-8 (from consortium SM-ER), respectively. a2 and a3 indicate TEM image and EDX analysis of the produced nano-particles by strain NTT-1, respectively. b2 and b3 indicate TEM image and EDX analysis of the produced nano-particles by strain SM-8, respectively. Bars indicate 50 μm (in a1 and b1) and 100 nm (in a2 and b2). Cu signals come from the mesh grid in a3 and b3.

Figure 13.10. PCR-DGGE analysis of microbial consortium SM-ER and the microbial isolates, strains SM-8 and SM-9. lane 1: SM-ER consortium incubated for one month, lane 2: SM-ER consortium incubated for two months, lane 3: strain SM-8 incubated for one month, lane 4: strain SM-9 incubated for one month. Forward primer (5′-GC clamp-CCTACGGGAGGCAGCAG-3′) corresponding to *E. coli* numbering position 341–357 (Bacteria V3 region), reverse primer (5′-CCGTCAATTCCTTTAAGTTT-3′) corresponding to *E. coli* numbering position 907–926 (Universal V5 region). 6% polyacrylamide gel (20%–80% gradient denaturing solution) was used for the electrophoresis at 200 V for four hours.

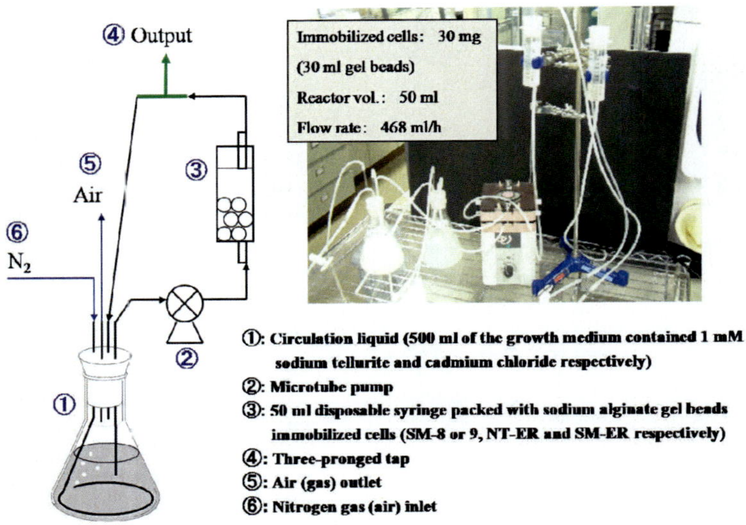

Figure 13.11. Bioreactor system for the simultaneous removal of Cd and Te ions based on conversion to CdTe particles.

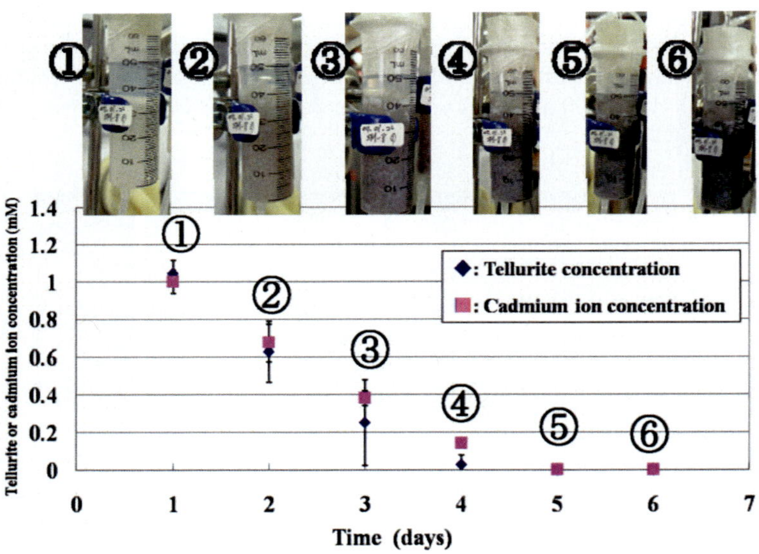

Figure 13.12. Appearance of bioreactor packed with sodium alginate gel beads that immobilized cells of the isolate (strain SM-8), and concentrations of tellurite and cadmium ion in the circulation liquid of the bioreactor system.

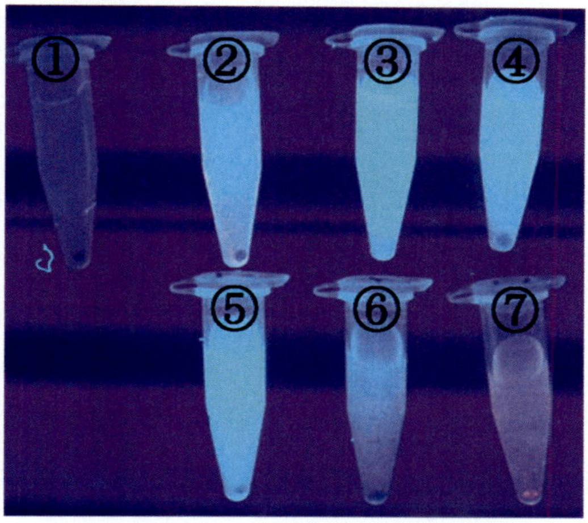

Figure 13.14. Light emission (lime green or blue fluorescence) from suspension of products generated by reductive combustion processing of collected cells in various enrichments. Cells were collected by centrifugation and then washed twice with distilled water. The collected cell paste was reductively burned at 300°C for six hours with ventilating nitrogen gas in a muffle furnace (Yamato FO 100). After burning the products were suspended with distilled water. 1: an enrichment from sediment of a pond in Imari, Saga, 2: a microbial consortium from Kinrin Lake sediment in Yufuin, Ooita, 3: SM-ER consortium (from sediment of an oil producing pond in Singleton memorial park in Kurokawa, Niigata, 4: an enrichment from a water way near Kameno river in Wakayama, 5: an enrichment from sediment of Arida river in Wakayama, 6: an enrichment from marine sediment of Osaka bay, Osaka, 7: control (*E. coli* JM109). UV at 365 nm was illuminated by a transilluminator (LX-200, TAITEC).

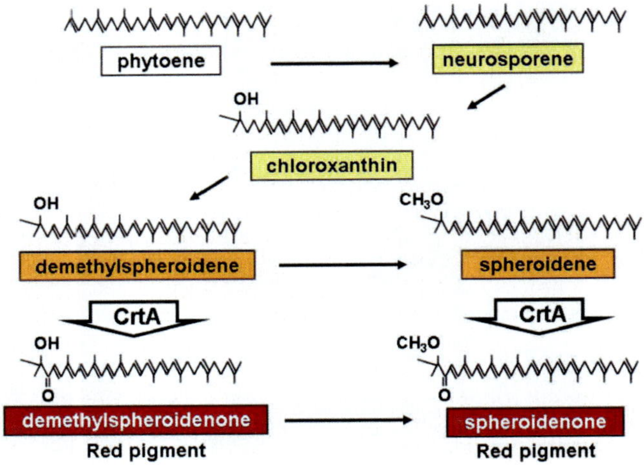

Figure 15.1. Carotenoid synthesis in spheroidene pathway.

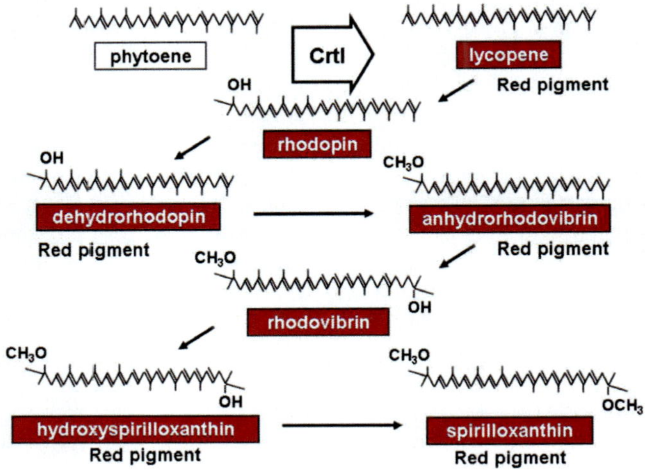

Figure 15.2. Carotenoid synthesis in spirilloxanthin pathway.

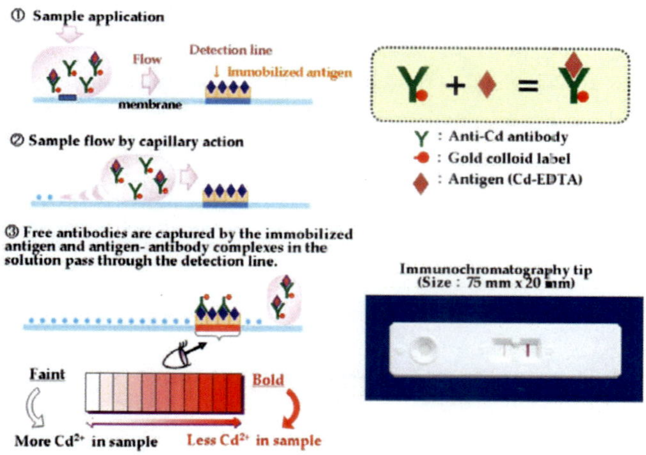

Figure 15.5. Immunochromatography for detection of cadmium.

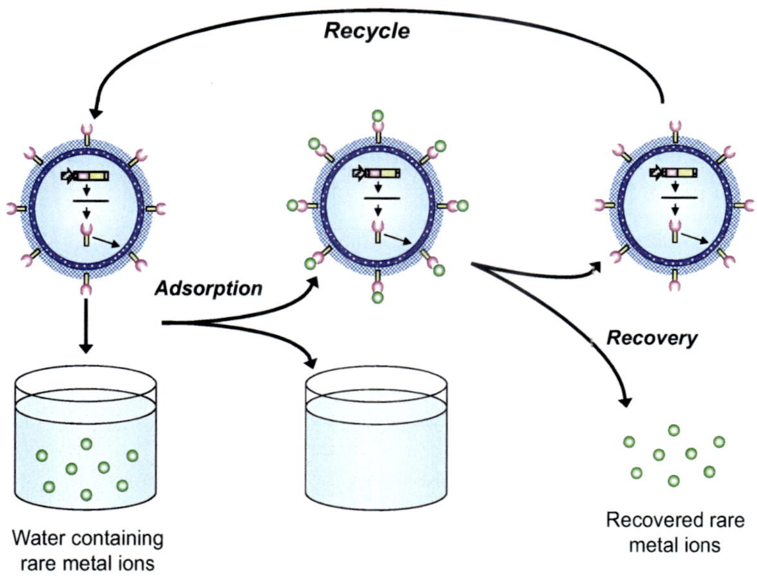

Figure 16.1. Adsorption, recovery, and recycling system of metal ions using cell-surface-engineered yeast.

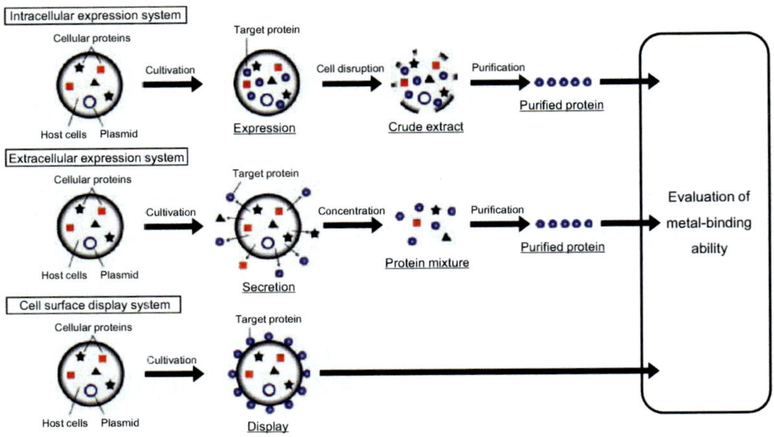

Figure 16.5. Cell surface display system as molecular tool for high-throughput screening of protein libraries.

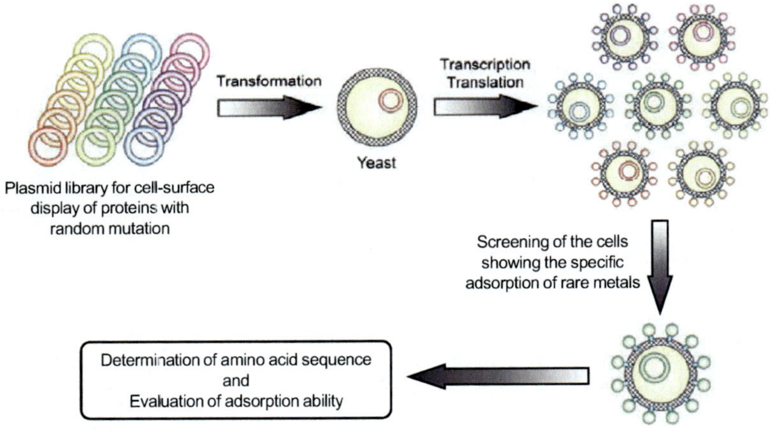

Figure 16.6. Creation of bioadsorbents that selectively adsorb rare metal ions by screening yeast libraries displaying peptides or proteins with combinatorial mutations.

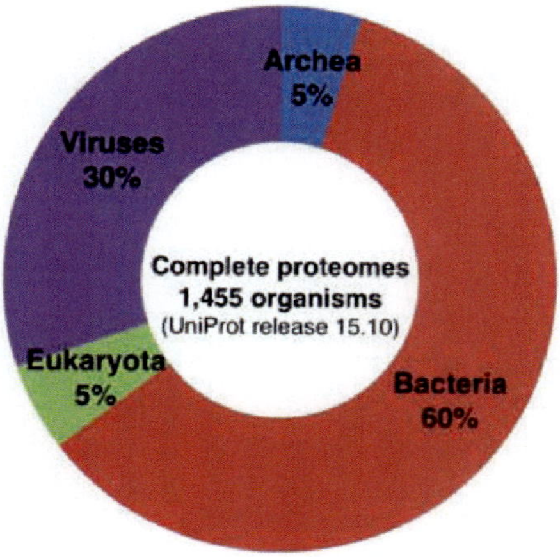

Figure 17.1. Distributions of organisms in the complete proteomes of UniProt. To be included in the complete proteomes, an organism must have a completely sequenced genome, i.e., fully closed and exhibiting either good gene prediction models or good quality transcriptome/proteome data. Therefore, for bacterial and archaeal genomes, whole-genome shotguns (WGS) and draft sequences are not included.